galà

Adventures of the
Most Well-Traveled **Filipinos**

DONALITO BALES JR.

2 January 2022

Dear Kevin,

Thank you for the friendship & our travel memories together. Inspire & be inspired!

Landon ☺

ISBN: 978-0-6452315-8-8

Edited by Reginald Yu
Book design and layout by Gram Telen
Published in Australia by Explore Next Level www.explorenextlevel.com

Table of Contents

PART 3: TRAVEL BUCKET LIST

PART 4: TRAVEL ADVICE

This book is dedicated to the memory of

Filipino World Travelers

Ryan de los Reyes, Meg Pipo, and Marcus Malabad

"If I have seen further, it is by standing upon the shoulders of giants." – Isaac Newton

Acknowledgments

Just as it takes a village to raise a child, so does writing a book. First and foremost, I would like to thank the 19 other Filipino world travelers whom I featured in this book for generously sharing their travel life experiences: Odette Ricasa, Jimmy Buhain, Luisa Yu, Riza Rasco, April Peregrino, Kach Howe, Raoul Bernal, Jon Opol, Jazz Gaite, Brian Banta, Auie Alipoon, Andie Novido, Henna Fuller, Kit Reyes, Rinell Banda, Salvador Casipit, Vhang Buechler, Ivan Henares, and Rambi Francisco. Collectively, we are the main characters of the book.

Thank you for the organizations that helped me find them from all over the world: Travelers' Century Club, Nomadmania founded by Harry Mitsidis, Every Passport Stamp Facebook Group founded by Stefan Krasowski, Most Traveled People founded by Charles Veley, and the Philippine Global Explorers global community of Filipino world travelers which Riza, Jimmy, Rambi and myself co-founded with Nolan Tianco and Mars Mendoza. Thank you to my travel heroes, Lee Abbamonte, Gunnar Garfors, Daniel Tirado, and Johnny Ward, whose travel writings provided the inspiration that fueled me to write this book.

Thank you to my sister Daryll Bales-Cale for connecting me to people who could share valuable insights to progress with self-publishing the book.

> *"None of us got to where we are alone. Whether the assistance we received was obvious or subtle, acknowledging someone's help is a big part of understanding the importance of saying thank you."* – *Harvey MacKay*

Thank you to my best friend Emee Hega for diligently proofreading and providing valuable critique on the manuscript to help ensure this book delights readers.

Thank you to the very talented Gram Telen for an awesome job making this book look like a stand-out bestseller – from the cover through to the interior visuals.

Thank you to my eloquent editor Reggie Yu for the impeccable editing effort from the early parts of this project through to its completion.

Thank you to my family and closest friends for their unwavering support in everything that I do, such as traveling the world many times over and writing a travel book about it.

Thank you to my loving parents, Danny and Nene, who gave me the best advice in my life: *"Go for your dreams and do whatever makes you happy."*

Thank you to the Creator for gifting me the ability to travel and the talent to co-create this book in sharing the abundance of creation. For the greater glory of God.

INTRODUCTION

'**Galà**' is a Filipino word for 'wandering around' or 'traveling in far places.' You are reading a one-of-a-kind travel book designed as a network of conversations. The cornerstone of this work rests on the collective experiences of people regarded as among the most well-traveled Filipinos in the world.

The book has eleven chapters divided into four parts. Intentionally-placed famous quotations are peppered throughout the book to highlight the content of that page.

Unlike most storybooks, this was NOT written to be read sequentially. One way you can read this book is to flip straight to the section that fits your particular need at the moment. If you want to be inspired by the lives of world travelers, go to Part 1. If you desire to be entertained by travel stories, go to Part 2. If you love to get bucket list ideas for your next trip, go to Part 3. If you like to learn helpful travel tips, go to Part 4. This book is a bit of an "organized chaos," much like traveling in the real world, so use it as you see fit on how it can best serve you.

Unlike most travel books, this is NOT a picture book. If you are expecting a photo album, this is not the title that you are after. However, the stories that were written here paint such a colorful canvas of rich experiences – the monochromatic swirl of sprawling mountainous vistas, the kaleidoscopic hues of countless customs and cultures, the somber shades of unfortunate occurrences that challenged each adventurer, and much more. The variegated narratives alone make it a matchless anthology which no picture book can hardly ever equal.

That said, you may wish to scan the QR code on the back cover that directs you to https://www.explorenextlevel.com/gala-book website, which has all the visual treats and exciting backstory extras that are sure to complement and enrich your experience while reading this book.

Part 1 (Traveler Profiles) introduces the biographies of twenty Filipino world travelers who have visited at least a hundred countries and territories, myself included. Most of us are county-counters who indulge in the practice of racking up the countries we see in our tally sheet, while some of us have lifelong missions of sojourning all the countries in the world. There are 193 official countries in the United Nations, but this list excludes countries like Taiwan and Kosovo. Several world travelers use 329, the number tracked by the Travelers' Century Club (TCC), an organization founded in 1954 for people who have visited 100 or more of the world's countries and territories. The book uses both numbers as terms of reference. Each traveler's accomplishments are up-to-date as of the 1st of August 2021.

A chapter is dedicated to nine Filipino travelers carrying Philippine passports, while another chapter features eleven Filipinos holding other credentials. I purposely made this distinction because it is more challenging to travel holding a Philippine passport, given the various restrictions imposed by foreign immigration, but this was never an obstacle for these adventurers bent on pursuing their dreams of travel. The same goes true for those coming from very humble beginnings; you'd be incredibly amazed to know that, from among our most well-traveled Filipinos, one served as a janitor, fast food service crew member, hotel front desk officer, gelatin vendor, jeepney driver, and

> *"The success of any great travel book can be
> measured not in awards, but in miles."*

even a beauty pageant contestant! Their successful journeys are but a shining testament to their resilience, as they gloriously overcame almost impossible odds to become the esteemed world traveler that they are today.

This book features quite a diverse group of travelers from a broad range of demographics in age, marital status, and professions. There are healthcare professionals, teachers, lawyers, engineers, I.T. professionals, entrepreneurs, travel bloggers, maritime professionals, travel agents, filmmakers, scientists, military personnel, and retirees. Our various life stories will inspire you, and I hope you can pick up some insights on how we ended up becoming world travelers that could be of value to you.

Part 2 (Travel Stories) has four chapters comprising of a smorgasbord of travel anecdotes ranging from exciting and happy chronicles, scary and frustrating accounts, funny and bizarre tales, as well as inspiring stories. Each section presents a series of thematic questions that collect selected responses from each of the twenty Filipino world travelers. The questions are:

- What are the most adventurous things that you have done while traveling?

- What's the craziest thing you've done in your travels?

- Have you done some trekking and hiking, and if so, which ones have been the most challenging?

- Describe your best travel buddy.

- What has been your most luxurious experience during your travels?

- How many photos do you take, and what has been the best photo you've shot?

- Did you get sick or hurt in an accident while traveling?

- Have you had a terrifying experience while traveling?

- What was your worst experience with customs, immigration, or border control?

- Describe your most uncomfortable transport ordeal.

- Describe the most stressful situation you've encountered while traveling.

- Were there any cultural practices that pushed your buttons?

- What's the funniest experience you've had while traveling?

- Describe a 'lost in translation' moment that you've had in your travels.

- Did you have some weird or surreal experiences?

- Is there any place in the world that you wouldn't dare go?

- Describe the best day ever during your travels.

- Describe the most interesting persons you have met while traveling.

- What is the most memorable local experience in your travels, and why?

- What is a place that has deeply moved you and why?

> *"I did not write half of what I saw, for I knew I would not be believed." – Marco Polo*

The fascinating responses from our peripatetic group will transport your imagination to an immersive recreation of what it's like to be a global traveler in a world of V.U.C.A.: volatility, uncertainty, complexity, and ambiguity.

Part 3 (Travel Bucket List) of the book is a good starting point for ideas on where to go and what to do based on carefully curated lists surveyed among the most well-traveled Filipinos. There is a chapter that showcases our highly recommended travel places which range from our top countries, cities, beaches, exotic locations, picturesque and artsy places, preferred locations to live, challenging destinations to visit, and places that make us exclaim, "Wow!"

Moreover, there are two sets of alternative lists of the Seven Wonders of the World and the Seven Wonders of Nature. And who is more well-placed to answer the question on the best Philippine destinations than a local. You can use this part of the book to help you plan your future trips, and you can even get to find out the jet setters' top bucket lists to give you further motivation and take your travel planning to the next level.

Then, there is also another chapter that provides travel experience highlights. You can tick off from many checklists, such as the top world festivals and events, as well as our highly recommended travel adventure activities that a number of us have done before. Our group of Filipino globetrotters would vouch for a list of favorite celebration parties, music festivals, world sporting events, wildlife experiences, culinary destinations, and party destinations. You will get a sense of being peregrine by finding out about our favorite world music, languages, world dishes, and even favorite spots to see beautiful people.

Part 4 (Travel Advice) is where you'll find valuable and practical bits of information that could help you prepare for your future trips. There are three chapters that focus on travel behavior, travel tips and tricks, and travel insights. In true crowdsourcing fashion, it has a similar style to Part Two, where there is a selection of responses from our cohort to the following set of questions:

- What travel-related question would you ask another traveler?
- How do you keep yourself entertained while traveling when you're not sightseeing?
- What is your traveling style (Do-It-Yourself, budget, luxurious, organized tour)?
- Describe some community service activities you were involved in while traveling.
- What's your favorite souvenir?
- What are some essential traveling skills you used?
- What are the important travel items you always bring with you?
- What are your tips for traveling on a budget?
- How much is your budget for your travels?
- How far in advance do you plan your trip?
- Which are the most valuable resources when planning your journey?
- What is the best learning you've gained while traveling?
- What were some of your rookie mistakes while traveling?

- What are the things you don't like about traveling?

- What inspires you to travel?

As a bonus, this section contains travel hacking tips, checklists for what to bring in a carry-on bag on international trips (for men and women), as well as a valuable list of travel resources, ranging from websites to mobile apps that the Filipino world travelers have used. I hope you will benefit a lot from the travel knowledge harvested through the wisdom of these adventurers.

Finally, this work has a similar message to the book published in 1870 by the first Filipino travel writer, Faustino Villafranca, who cited that he expected it "to be useful for the people of my country who want to know the world." My intention is also to have the rest of the world understand our shared humanity through the experiences of the Filipino world travelers captured here. After all, a common thread runs across each page of this book: our ongoing love of travel. It is undeniable that the most well-traveled Filipinos possess the DRD4-7R 'wanderlust' gene. And you likely have that too if you're reading this book, so welcome to our tribe!

I wrote this book when world travel has become very challenging due to the COVID-19 pandemic. Hence, I hope this will plant a seed for re-igniting a post-pandemic travel industry once it's safe to travel the world again.

"I beg young people to travel. If you don't have a passport, get one. Take a summer, get a backpack and go to Delhi, go to Saigon, go to Bangkok, go to Kenya. Have your mind blown. Eat interesting food. Dig some interesting people. Have an adventure. Be careful. Come back and you're going to see your country differently, you're going to see your president differently, no matter who it is. Music, culture, food, water. Your showers will become shorter. You're going to get a sense of what globalization looks like. It's not what Tom Friedman writes about; I'm sorry. You're going to see that global climate change is very real. And that for some people, their day consists of walking 12 miles for four buckets of water. And so there are lessons that you can't get out of a book that are waiting for you at the other end of that flight. A lot of people—Americans and Europeans—come back and go, Ohhhhh. And the light bulb goes on." – Henry Rollins, Punk Rock World Traveler

Whether you are a seasoned globetrotter or an aspiring itinerant, this is a legacy work crafted with much love that hopes to grant you the gift of inspiration to discover the world through your own eyes. Allow the gift to find you, consume you, then give it away.
— **Donalito Bales Jr.**

PART 1:

TRAVELER PROFILES

CHAPTER 1

FILIPINO WORLD TRAVELERS USING PHILIPPINE PASSPORTS ONLY

Dominador "Jimmy" Buhain

Bwindi Impenetrable National Park, Uganda

- **Philippine Hometown:** Quezon City, Metro Manila
- **Primary Residence:** Quezon City, Philippines
- **Profession:** Lawyer and Businessman
- **Age:** 76
- **Traveled Continents:** 7
- **Traveled Countries & Territories:** 248 (158 U.N.)
- **Travel Site:** https://www.ddbuhain.com

Jimmy is the second-generation patriarch of the Buhain and Fontelera families, whose business of bookselling, publishing, and printing educational materials have since become a national institution, after having begun as a second-hand book dealer and importer more than seventy years ago. He has certainly gone a long way from his humble beginnings, having been born to middle-income parents in Tondo, Manila. He studied at San Beda University during elementary and high school and later completed a Bachelor of Science degree in Business Administration at the De La Salle University. Jimmy initially pursued his law studies at the Ateneo Law School but then transferred to Far Eastern University, where he earned his Bachelor of Laws degree and subsequently passed the bar in 1977. He was one of the first graduates of the Mandatory Continuing Legal Education at the University of the Philippines College of Law in 1978.

A multi-awarded businessman and civic leader, he was bestowed with the 2017 "Distinguished Golden Tamaraw" Award from the Far Eastern University. He received the "Hall of Fame" Award in 2013 from San Beda, the "Icon of Entrepreneurship" Award in 2014, as well as the Most Distinguished Bedan Award, and the Bedan Achiever Award. He was the President of the San Beda University Alumni Association for five consecutive years. He was also Chairman of the Board of Trustees of La Consolacion Manila, Chairman of La Consolacion University of the Philippines in Malolos, Bulacan, and Honorary Chairman of La Consolacion of Biñan, Laguna and Tanauan, Batangas. Other key civic responsibilities he took on was being Past President of Club Bulakeño for two years and Past President of the Ayala Heights Village Association for six years.

Jimmy has established himself as a well-respected businessman being the Chairman and President of the Rex Group of Companies, one of the leading Philippine conglomerates of the book industry. In 2018, he was given the Outstanding CEO Award by the Philippine Council of Deans and Educators in Business. Having served as the first salesman of their family business, he was privileged to visit all the provinces in the Philippines, which eventually led to expanding Rex Bookstore to twenty-five branches all over the country.

On June 12, 2006, Jimmy completed traveling to all Philippine provinces, arguably making him the first Filipino to do so.

Jimmy served as President in a number of key organizations in the book industry, such as the Asia-Pacific Publishers Association, the ASEAN Book Publishers Association, the Philippine Printing Technical Foundation, a Forum for Asia Pacific Graphic Arts, Philippine Book Publishing Development Federation, and the Printing Industry Board Foundation, Inc. He was elected President Emeritus of the Philippine Educational Publishers Association and became Founding Chairman of the National Book Development Board. He co-authored a number of books entitled *"History of Publishing in the Philippines," "History of Printing in the Philippines,"* and *"The Survival of Books, Book Stores and Other Display Areas in the Philippines."*

From his relatively modest childhood, Jimmy began the desire to travel when he saw the movie *Peter Pan* who was gifted with the ability to fly. He realized early on that he will not settle for being just a well-wisher; instead, he yearned to be the one who would be sent off to fly.

As the family bookstore business improved when they shifted from buying and selling second-hand books into importation of books from abroad and venturing into book publication, his parents gave him the opportunity to travel overseas and visit Singapore, Hongkong, and Japan. Because of the beauty and grandeur of what he had seen and experienced during those trips, he resolved that he would save, work hard and take care of his health to see more destinations. He learned early on that the three elements that need to come together to enable travel are money, time, and health.

Incidentally, Jimmy's formal first name is *Dominador* which means "to conquer" in Spanish. Over the last few decades, he traveled around the world. He conquered the obstacles of traveling to so many countries, using only his Philippine passport, and immersing himself in diverse cultures. His elder brother encouraged him to join the Travelers' Century Club (TCC) once he reached one hundred countries and territories, the TCC entry criteria for membership.

Just before the pandemic lockdown restrictions were imposed in March 2020, Jimmy's count was already 248, which was short of two more countries in order to qualify for the TCC's Platinum status. In June 2010, he was recognized as the "Most Traveled Person" Award in the Asia-Pacific by the MTP (Most Traveled People) Extreme Travel Club. In July 2021, Jimmy became the first Filipino on NomadMania to have been verified for NomadMania regions and UN countries. Through this network of world travelers, he managed to connect with other Filipino globetrotters, which eventually led to the formation of the Philippine Global Explorers, of which he is a co-founder and serves as the Founding Vice Chairman and Charter Secretary. As of 2021, Jimmy has the distinction of being the Filipino to have visited the most number of countries and territories using only a Philippine passport.

Having been to 408 world heritage sites, Jimmy has accumulated many souvenirs representing the countries he visited which doubles as trophies of his journeys. He set up several museums to display these souvenirs, which can be found at: (1) Book Museum-*cum*-Ethnology Centers in Marikina City; (2) Rex Display Area in Tanza, Cavite; and the (3) Banaue Heritage Hotel and Museum in Ifugao, which has a display of artifacts from the Cordilleras. Some of his other souvenirs are displayed in his Rex office in Quezon City. These endeavors inspired him to establish his Foundation whose purpose is to collect artworks and other cultural items and showcase the diversity of the culture of each country or territory, with the earnest hope that it would bring bridges of understanding that would lead to peace, harmony, and prosperity to everyone: Filipinos and non-Filipinos alike.

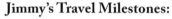

"We've always defined ourselves by the ability to overcome the impossible. And we count these moments. These moments when we dare to aim higher, to break barriers, to reach for the stars, to make the unknown known. We count these moments as our proudest achievements. But we lost all that. Or perhaps we've just forgotten that we are still pioneers. And we've barely begun. And that our greatest accomplishments cannot be behind us, because our destiny lies above us." – Cooper, from the film Interstellar

Jimmy's Travel Milestones:

✓ Traveled 248 out of 329 countries and territories (159 out of 193 U.N. countries)

✓ Reached 100th country & territory in 1999, 150th in 2002, and 200th in 2008

✓ Visited 7th continent in 2003

✓ Completed trips to all UN countries in Europe in 2018

✓ Visited 408 UNESCO World Heritage Sites

Katrina "Kach" Umandap-Howe

Moai statues at Easter Island, Chile

📍 **Philippine Hometown:** San Pablo City, Laguna

🏠 **Primary Residence:** Herceg Novi, Montenegro

📷 **Profession:** Travel Blogger

😊 **Age:** 32

✈ **Traveled Continents:** 7

🌐 **Traveled Countries & Territories:** 168 (146 U.N.)

🌐 **Travel Sites:** https://twomonkeystravelgroup.com
https://mrandmrshowe.com

Kach was born in Makati City, Philippines, where her father's family is based. She and her two other siblings were raised by her mother and grandparents in San Pablo City, Laguna, having grown up in a broken family. She has always been a hard worker and found ways to support herself financially even as early as elementary school. During her high school year, she became the student council president and was the representative of their city for the *Gawad Felicisimo T. San Luis Para Sa Namumukod Tanging Kabataan ng Laguna* (Search for Outstanding Youth of Laguna). Kach went on to study at the University of the Philippines in Los Baños and graduated with an economics degree in 2009. While in college, she was an active leader in a number of extra-curricular activities such as the U.P. Los Baños Economics Society, where she organized concerts and fundraising events, as well as having arranged one of the biggest car shows in Laguna as a member of U.P. Los Baños Sigma Beta Sorority.

The first time Kach traveled outside the Philippines was a month after her college graduation at twenty years old. She flew to Kuwait to do on-the-job training work at the Philippine Overseas Labor Office. She originally planned to do it only for three months before going back to the Philippines to take up law and pursue her dream of becoming a diplomat. Eventually, she ended up living in Kuwait for the next three years, where she was employed in a dental company before becoming a Quality Assurance Officer and Executive Assistant to the Chief Operating Officer. Kach later changed jobs and assumed the lofty position of "Quality Assurance Supervisor on Occupational Safety and Health Policy-Making" for an internationally-renowned hospital. Kach moved to Erbil, Kurdistan, in Iraq in 2012 to continue working for an oil company. Despite this financial success for a twenty-four-year-old Filipina, she still felt a void in her life. Her family, peers, and colleagues told her she was crazy to waste the opportunities that she had at that time. But Kach followed her heart, and she eventually quit her nine-to-five career to pursue her travel passions. She headed off backpacking to Southeast Asia in 2013, bringing her siblings along so they can experience their first overseas travel.

Kach ended up teaching English in Vietnam for seven months. She then flew to India and studied to become a Tantra Yoga teacher and Ayurveda massage therapist. She did this while working in various odd jobs, as a cook, hostel cleaner, and waitress. In late 2014, having embraced a nomadic lifestyle, she began travel blogging on a full-time basis. After four years of adventure traveling followed by two years of sailing the Caribbean, she moved and bought a stone house villa in Herceg Novi, Montenegro, in 2019 to start a new expat life.

Since starting her own online travel business, Kach has been self-employed. Through her travel blog, she's worked with airlines, tourism boards, luxury hotels, big tour operators, start-up companies and sponsors that have helped her travel around the world using her Philippine passport. She bought her first house, mortgage-free, from her travel blogging income. She's very proud to have been able to bring her mom to different countries every year since 2015 including, joining a Caribbean cruise, a luxury trip to New York, and visiting countries like the United Kingdom, Jordan, and Morocco. Kach loves traveling solo, which she did when she traveled to 16 African countries, then Eastern Europe, and an overland trip all over Central and South Asia.

Her travel blogging success story has been featured in Forbes, TIME. com, Daily Mail, Business Insider, NYTimes, Huffington Post, and Aljazeera News, among others. Her travel blog has been nominated and won various travel blogging awards in the United States of America, United Kingdom, Philippines, and South Africa. Kach has been a staple in the speaking circuit to such travel events as the World Tourism Forum in Istanbul, Turkey, in 2017, among many others.

Always striving to move forward and learn something new, Kach vividly remembers the time she learned how to drive a motorbike in the chaotic city of Hanoi, Vietnam. She's learned to embrace her fear and do it anyway, as she has done with such extreme sports as skydiving, bungee jumping, and paragliding despite suffering from acrophobia. She employed the same courage that led her to leave her stable corporate job for an unknown frontier of traveling, which she never regretted. She may not have the same luxury and security she had in the corporate life, but she found something more fulfilling — the freedom and inner happiness that she was searching for all her life.

Kach aims to travel to every country in the world using only her Philippine passport. The law of attraction is her life's philosophy as she believes that what happens to her is the result of her attitude. She believes that only when you try to visualize your dreams will these be converted to words and actions (encapsulated as your goals), then you will start attracting the very things you wish. This has been working for her as she fittingly expresses in these inspiring words, "So, to my fellow women, Filipinos, and backpackers... yes, you can have the traveling life you want, earn money and be successful at the same time."

Kach's Travel Milestones:

✓ Traveled 168 out of 329 countries and territories (146 out of 193 U.N. countries)

✓ Reached 100th country and territory in 2017, and 150th in 2018

✓ Visited 7th continent in 2017

✓ Completed trips to all UN countries in Southeast Asia in 2017, South/Central Asia in 2019

✓ Completed trips to all new 7 Wonders of the World in 2019

Joniper "Jon" Opol

Pico de Loro, Batangas, Philippines

📍 **Philippine Hometown:** Davao City

🏠 **Primary Residence:** Metro Manila, Philippines

💼 **Profession:** I.T. Professional

🕐 **Age:** 39

✈ **Traveled Continents:** 6

🌐 **Traveled Countries & Territories:** 162 (141 U.N.)

🌐 **Travel Site:** https://www.instagram.com/wanderjon/

Jon was born in Davao City, where he grew up loving history and geography from an early age. He had memorized the world map, every country's capital, flag, currency, form of government before he even turned ten. *The Atlas* and the *Guinness World of Records* books were among his refuge. He started collecting stamps and coins from places around the world. His upbringing and his love for reading exposed him to the vast world beyond the confines of the humble village where he grew up. He already knew he would see the world even in his youth, and it was just a matter of when.

Jon finished his secondary education at Salaman Institute in Sultan Kudarat, Mindanao, and then completed a degree in Accountancy from the University of the Philippines. He then worked as an I.T. professional specializing in Finance enterprise resource planning systems for companies in various industries ranging from food & beverage, oil, and banking.

In late 2007, Jon had his very first trip abroad for a conference in Hong Kong. At twenty-three, he was emboldened entirely with the experience from that first trip that he tendered his resignation the following week to his employer in the Philippines and left the country for good. Luckily, he immediately got a job in Singapore, which served as his springboard to further opportunities. He felt it was the wisest decision he has ever made and was a risk that really paid off.

He got relocated by his company to work in different countries where his clients were based. Because of this, he was very mobile and became peripatetic in nature. He ended up living in Australia,

Dominican Republic, Germany, Indonesia, Italy, Netherlands, Singapore, South Korea and the Unites States for at least half-a-year in each country. Whenever he is in a new place, Jon would explore as much as he could after seeing everything the new city can offer. He would travel as far as his budget can afford and as much as his time can accommodate. He has also stayed with friends all over the world that he connected with through Couchsurfing. Jon considers himself an intrepid traveler who likes exploring new stuff and enjoying meeting new friends. He has a mission to visit every country in the world.

Jon had to quit his work several times to embark on months and even year-long backpacking. He prefers modern, vibrant, clean, rich cities with cultural and gastronomic hotspots, ideally places with easy access to outdoor activities, especially mountains. One of the happiest events in his life was when he quit his work in 2009 and did a 5-months backpacking trip from Turkey all the way to Egypt, United Arab Emirates, Nepal, and India. He never looked back after that experience. He did the same thing quitting his job to travel in 2013, then in 2016, and then in 2019. He has taken at least 500 flights in his lifetime. In 2018, he was recognized as one of the Star Alliance's "Top 1,000 Travelers of the Year."

Aside from traveling, Jon likes to read, watch sporting events, play tennis, hike and run. He loves running so much that everywhere he goes, he brings along his running gear. He would like to run as many full marathons at every place he visits. He has completed seventeen full marathons across five continents, including the New York City Marathon, Chicago Marathon, and Berlin Marathon.

Jon also loves photography. As of 2021, he has posted around 2,000 images that he shared with his approximately 18,000 Instagram followers. He became the first Facebook group moderator of the Philippine Global Explorers travel community.

His personal philosophy in life is to follow the tenets of simplicity, kindness, ambition, and generosity.

Jon's Travel Milestones:

✓ Traveled 162 out of 329 countries and territories (141 out of 193 U.N. countries)

✓ Reached 100th country and territory in 2017, and 150th in 2019

✓ Visited 6th continent in 2013

✓ Completed trips to all new 7 Wonders of the World in 2016

Andros "Andie" Novido

Antarctica

- **Philippine Hometown:** Tondo, Manila
- **Primary Residence:** Quezon City, Philippines
- **Profession:** Aircraft Maintenance Planning Engineer
- **Age:** 41
- **Traveled Continents:** 7
- **Traveled Countries & Territories:** 145 (127 U.N.)
- **Travel Site:** https://www.theviewingdeck.com

> *"The life you had led doesn't need to be the only*
> *life you have." – Anna Quindlen*

Andie was born into a low-income family in Tondo, Manila. His father works as a mechanic, and his mother is a housewife. They lived together with his two other siblings. He worked as a gelatin vendor when he was in primary school to augment his family's meager income and saved on daily transportation expenses by walking from his home to school, a distance of five kilometers, every day for six years. Determined to give a better life for his family, he studied hard, eventually ranking Number Nine of his graduating class of more than 1,000 students in Tondo High School.

He was later employed as a part-time service crew of a popular fast-food chain for three years while pursuing his Bachelor of Science degree in Electronics and Communications Engineering at the Polytechnic University of the Philippines. He later passed the Philippine Civil Service Exam and the government-administered Electronics and Communications Engineering Licensure board examinations, despite not having attended formal review classes and working as a regular IT staff on a graveyard shift. He studied his Master of Engineering Degree at De La Salle University as a resident university scholar while working as a regular employee. He has worked for more than thirteen years in the commercial aircraft maintenance industry, having been employed in a number of airline engineering support companies serving Lufthansa Technik, Kuwait Airways Production Planning, and Etihad Engineering. He later worked as an Overseas Filipino Worker in the Middle East for eight years as an Aircraft Maintenance Planning Engineer. Within just two years working abroad, he was able to purchase his own fully-renovated house and lot in Metro Manila. Andie's love affair with traveling began when he was about to complete his graduate studies in 2010. Looking for a hobby to kill his boredom, he decided to muster the courage to undertake a solo backpacking adventure as a fun-hiker and later, as a budget trip organizer and a travel companion, in an attempt to explore the world.

Andie's travel personality is pursuing minimalistic, do-it-yourself budget trips. He carries only a lightweight backpack as much as possible and makes it a point never to check in his luggage during his trips. When visiting a place, he sticks to his planned itinerary so he doesn't waste time, ensuring to include the major tourist spots, yet willing to try things that more tentative travelers might skip. He uses his weekends and holidays from work to live out his fast-paced travel lifestyle.

As a self-confessed individualist, he travels almost always alone. He did, however, go backpacking with his sexagenarian mother on trips to Hong Kong, Macau, Japan, United Arab Emirates, Spain, Andorra, Portugal, and Italy. He even accompanied his parents, siblings, and nephews to the United Arab Emirates for a month as tourists. At one point, Andie resigned from his high-paying job to fulfill his dream of doing a self-guided solo expedition at the acclaimed hiking routes of the Himalayas.

As a nature-lover and adventure-seeker, he is an ardent hiker, biker, sightseer, and landscape photographer and has embarked on more

> *"To travel far, far – and that first morning's awakening*
> *under a new sky! And to find oneself in it – no, to discover*
> *more of oneself there." – Rainer Maria Rilke*

than ninety solo trips, including solitary hiking forays at the Himalayas in Nepal, Jebel Shams in Oman, Mount Bisoke, Mount Ramelau in Timor Leste, Table Mountain in South Africa, Ala Archa National Park in Kyrgyzstan, Patagonia in Argentina, Tajumulco Volcano in Guatemala (Central America's highest peak), Paricutin Volcano in Mexico (World's Youngest Volcano) among many others. His longest hike in one day was a twenty-two-hour brutal traverse trail in Mount Mariveles in the Philippines; the highest mountain altitude gained was around 2,000-meter ascent to Tungurahua Volcano Refuge Camp (3,830masl) in Ecuador. His longest bike ride was a nine-hour cycling trip in Abu Dhabi, which spans one-hundred-and-fifty kilometers using only a folding bike.

For someone who does not know how to swim, cannot drive a car, and suffers from acrophobia (fear of heights) and tachophobia (fear of speed), Andie managed to accomplish the unbelievable: a remarkably long bucket list of global adventures. Andie completed a seven months Latin America solo trip from Mexico down to Central and South America visiting 16 countries from November 2020 to June 2021. In addition, he was the only non-Arab out of eight participants in the 2013 AXE Apollo Space Academy Middle East Grand Finale to compete for a once-in-a-lifetime opportunity to fly aboard the suborbital space plane Lynx.

In January 2020, Andie circumnavigated the world in a two-week solo trip, traveling from UAE to Chile, crossing to the United States, then to the Philippines, then navigating back to UAE with multiple stops in Easter Island, Uruguay, Peru and South Korea. Moreover, he took the longest, non-stop train ride in the world for more than six days, via Russia's Trans-Siberian Railway from Moscow to Vladivostok, passing through seven time zones.

Throughout his global travels, Andie has been to a number of contrasting environments: the coldest place he's ever been was in Alaska, USA, with a biting temperature of -36'C; while the hottest was a sizzling +52'C in Kuwait; the remotest place he's reached in the north was in Fairbanks, Alaska, USA, while the farthermost south was in the Antarctica Peninsula; the shore of the Dead Sea in Israel was the lowest point on dry land he's ever gotten, and the highest was at the Kala Pathar Peak and the Everest Base Camp in the Himalayas.

Outside of travel, Andie's other passions include drinking local beers, environment and wildlife protection, volunteering activities, and doing random acts of kindness to local strangers. The best advice he'd like to share with others? *"Live a life that you will always remember."*

Andie's Travel Milestones:

✓ Traveled 145 out of 329 countries and territories (127 out of 193 U.N. countries)

✓ Reached 100th country & territory in 2019

✓ Visited 7th continent in 2019

✓ Completed trips to all UN countries in Southeast Asia in 2018, South/Central/North Asia in 2019, and Central America in 2021

✓ Circumnavigated around the world in a 2 weeks solo trip in 2020

✓ Completed trips to all new 7 Wonders of the World in 2021

Aulorence "Auie" Alipoon

Deira Old Souk Abra Station, Dubai, United Arab Emirates

📍 **Philippine Hometown:** Olongapo, Zambales

🏠 **Primary Residence:** San Diego, California, USA

💼 **Profession:** Entrepreneur

📅 **Age:** 37

✈ **Traveled Continents:** 7

🗺 **Traveled Countries & Territories:** 137 (109 U.N.)

🌐 **Travel Site:** https://www.facebook.com/Auiee

https://www.instagram.com/Aulorence/

Auie was born in Olongapo City and took up her elementary and high school years in St. Joseph College-Olongapo (formerly St. Joseph's School). She subsequently took a Computer Engineering course at AMA College while working part-time at Legenda Hotel casino in Olongapo City. Determined to raise her academic profile, she pursued her Master in Business Administration at AMA University in 2001. Naturally statuesque and attractive at 5'9, she represented her city in a number of prestigious beauty pageants, including the Bergamo's Modelquest in 2001, Mutya ng Pilipinas in 2002, and Binibining Pilipinas in 2005.

Apart from her extensive work experience, she marshaled her innate wit and charm to land a lucrative job in a U.S.-based casino company, the Carnival Corporation Casino Division, in 2007. Within three months from her hiring, she quickly rose through the ranks, becoming a casino executive, largely through hard work and perseverance. This was her ticket to travel around the world, a long way from her first overseas trip to Palau way back in the nineties. She would normally work six months, traveling around the world, and get another two months of vacation leave after that which she used to travel some more. By 2010, she had circumnavigated around the globe; by 2011, she had reached her seventh continent. She is an avid photography buff and would have amassed an extensive collection of travel photos had her hard drive containing a treasure-trove full of memories not crashed.

Auie has not been spared of personal crises. Her mother's sudden death in 2015 due to aneurysm left her devastated. Consumed with grief, she found comfort in food and gorged her way to a

> *"We travel for romance, we travel for architecture,*
> *and we travel to be lost." – Ray Bradbury*

whopping 100kg. In a moment of epiphany, she realized that she needed to move forward and was determined to regain her self-confidence. She became a champion of keto low-carb program, lost over 36 kg (80 pounds) in a span of 8 months, and inspired thousands with her weight-loss journey. In October 2019, she was invited by the world-renowned fashion designer Rocky Gathercole to model his creations during the Philippine Fashion Week.

She also rekindled her love for traveling and began to pursue more adventurous itineraries, particularly avoiding urban concrete jungles. A self-confessed dog-mom and a cynophile, she regrets leaving her dog during trips, comforted by the fact that she would be able to explore nature and scenic places around the globe. Some of Auie's memorable trips include those in Antarctica, Cannibal Village in Papua New Guinea, Bora Bora in Tahiti, Tromsø in Norway, La Goulette in Tunisia, Seychelles, Madagascar, Nicaragua, Falkland Islands, Bermuda, Aruba, and the American Samoa, as well as her backpacking adventure all over Europe.

After a colorful thirteen years as a casino executive, she decided to retire in June 2019 to focus on her personal life. She relocated to San Diego, California, USA, in March 2020 to begin her new chapter as an entrepreneur, managing rental property investments and doing digital marketing initiatives. This self-proclaimed bachelorette is now ready to find herself a lifetime partner who would perfectly complement her self-made, alpha-female ambivert self. For now, she'll be taking a break from the lure of world travel in exchange for finding love, getting married, and having a family… in the hopes of traveling one day again, no longer alone, but this time, with her loved one.

Auie's Travel Milestones:

✓ Traveled 137 out of 329 countries and territories (109 out of 193 U.N. countries)

✓ Reached 100th country and territory in 2017

✓ Visited 7th continent in 2011

✓ Circumnavigated around the world from January 2010 to May 2010

Rinell Banda

Northwest Passage, Arctic Circle

📍 **Philippine Hometown:** Las Piñas, Metro Manila

🏠 **Primary Residence:** Parañaque City, Philippines

📷 **Profession:** Youtube Vlogger / Entrepreneur / Retired Seafarer

✈ **Traveled Continents:** 7

🌐 **Traveled Countries & Territories:** 125 (125 U.N.)

🌐 **Travel Sites:** https://www.youtube.com/c/BuhaysaCruiseShip/

https://www.youtube.com/user/rinellbanda

Rinell has always dreamed of traveling the world. As a kid, he wanted to become a pilot to achieve his dream to travel. However, due to the prohibitive costs of pilot school and his personal financial constraints, he wasn't able to study how to fly planes. He did manage to graduate as an aircraft mechanic and completed a Bachelor of Science degree in airline business administration from the PATTS College of Aeronautics.

After graduating from college, he accepted a number of odd jobs to make ends meet. He worked as a service crew worker for several fast-food restaurants (i.e., Jollibee, Carl's Jr., Dunkin Donuts); he studied a course on becoming a waiter, bartender, and housekeeper, which led him to work as a banquet waiter in a Makati City-based, five-star hotel for a year. Little did he realize that this was his stepping stone towards achieving his lifelong dream of traveling the world. Through his hotel and restaurant management training, he found his way in the maritime waters, as he finally got a good break through Magsaysay Maritime Corporation, where he worked at Costa Cruises as a seafarer in 2008.

However, life threw him another curveball. He got sick on one of these trips and had to be medically repatriated back to the Philippines. As a way of trying to recover his dwindling savings, he ventured into more odd jobs, finding work as a sales agent, a fitness instructor, a massage therapist, a commercial model, and as an entertainer for a few short months as part of German Moreno's "Walang Tulugan with

> *"Never give up on a dream just because of the time it will take to accomplish it. The time will pass anyway." – Earl Nightingale*

the Master Showman" TV show in Japan. Rinell's prayers were answered under what he believes was God's perfect timing, as he found an opportunity to work again in the cruise industry. He found a job as a marine crew for MV Artemis and, later, The World, a private mega-yacht that happens to be the world's largest privately-owned cruise vessel that circumnavigates the globe every two to three years. After working as a deck steward, he rose through the ranks, being promoted to bellman and later, as bell captain for the yacht. He also learned how to drive a zodiac boat which became a lucrative sideline during expeditions. This gave him the opportunity to travel around the world extensively for free.

Apart from his laser-like determination to succeed, what kept Rinell going was his love for his family. Because he missed them a lot being out at sea for a long period of time and wanted to find a way to be able to share updates of his life back with them, he started making videos, using his digital camera, tripod, and drone to document the countries that he visited, capture the picturesque views of the places, interview locals, eat street food, as well as sing and dance on the streets. His channel became so popular on YouTube that he was inspired to open his own Pinoy Travel Channel in 2011, where he came to be known as "Dronie Man."

By April 2014, he has created over 100 videos after visiting 88 countries, some of which were clips christened "Where on Earth is Rinell?" that showed him dancing with people around the world. This got him featured in the Philippine TV show "Kapuso Mo, Jessica Soho" as a Filipino world traveler who continued to pursue his dream despite his setbacks in life.

In his desire to mentor aspiring cruise ship professionals to have a thriving career in the cruise industry, he created the "Buhay sa Cruise Ship" video blog and produced a Youtube channel by the same name. As of 2021, the "Buhay Sa Cruise Ship" channel has over four million views and exceeded 45,000 subscribers. Corollary to this, he created the "Buhay sa Cruise Ship" Facebook Group seafaring community in 2017, which has grown over 350,000 members as of 2021.

After amassing a sizable fortune during his years in the cruising industry for over a decade, Rinell diversified his financial investments to accelerate his retirement by the end of 2019, so he can wholly focus on being an entrepreneur, support the seafaring community he founded, and take care of his one-year-old baby daughter.

Rinell believes that a combination of dedication, hard work, determination, humility, and prayers were the key elements to his success. And that's how he got to travel the world for free.

Rinell's Travel Milestones:

- ✓ Traveled 125 out of 329 countries and territories (125 out of 193 U.N. countries)
- ✓ Reached 100th country and territory in 2016
- ✓ Visited 7th continent in 2012
- ✓ Completed trips to all new 7 Wonders of the World in 2014
- ✓ Circumnavigated around the world in 2019

Salvador "Badong" Casipit

Luján Zoo Reserve, Buenos Aires, Argentina

📍 **Philippine Hometown:** Pangasinan

🏠 **Primary Residence:** San Jose Del Monte, Bulacan, Philippines

📷 **Profession:** Maritime Professional

📅 **Age:** 47

✈️ **Traveled Continents:** 7

🗺️ **Traveled Countries & Territories:** 120 (105 U.N.)

🌐 **Travel Sites:** https://www.facebook.com/cheapatonictraveler
https://www.facebook.com/DiscoverAdventureTravelExplore

Badong was born in Quezon City to a family with three other siblings. His father is a family driver, and his mother is a housewife, both from Pangasinan. As a child, his mother brought him frequently to their hometown of Infanta, where he wandered around the nearby seas and mountains with his cousins and friends. He studied elementary in Barangay Bayambang in Infanta, Pangasinan. As a Grade One pupil, he used to accompany his grandfather to his small farm in the mountains, where he learned many things about nature and survival, despite having a fear of heights. After finishing high school, he took a two-year vocational course in computer programming at STI College.

Badong worked on a number of jobs to support his studies, which included working as a dishwasher inside the Bangko Central ng Pilipinas employees' canteen. He worked as a janitor at the Department of Environment and Natural Resources main office in Quezon City. He also worked in a marble factory as a storekeeper. As a sideline, he worked as a cigarette and newspaper vendor in Quezon City around the Delta, Circle Theater, and Timog area.

He worked for about four years as a family driver, eventually as a taxi driver and then as a jeepney driver. His best friend's parents taught him to bake, so he became a seasonal baker making fruit cakes during Christmas. This helped to support his vocational course on dressmaking which he later pursued. Having learned to cut and sew formal dresses, he worked as a garment factory supervisor overseeing quality control processes for various well-known brands. Being the breadwinner of his family, he worked hard to be able to support them.

*"When something is important enough, you do it even
if the odds are not in your favor." – Elon Musk*

His love for traveling led him to become a professional maritime seafarer. He got a job at Premier Cruise Line in his early twenties, where he got to visit the United States, his first country outside of the Philippines. Originally hired as the ship's in-house tailor, he ended up working as a galley utility responsible for cleaning dishes. During his first week on the job, he wanted to quit as he found that work very challenging, as he had to clean large food service equipment and utensils. Fortunately, his request to transfer to Provision as a utility storekeeper was granted; there, he managed to exercise his computer skills to render administrative work, track inventory, and issue stocks to the galley. It was at this time that he traveled extensively in the Caribbean, which was the route of the "Big Red Boat," the cruise ship that he was working at that time.

When the Premier Cruise Line closed shop, Badong was called to work in P&O Cruises in 2001. A few years later, in 2004, he got an opportunity to travel on a world cruise twice; it sailed from the United Kingdom to Australia and back through the Indian Ocean, passing through numerous cities in Europe, Africa, Middle East, Asia, and Australia. He was able to do this again in 2006, in both instances, using his Philippine passport while working on the cruise ship. It was during this trip that he realized that his desire was consumed by traveling.

He encouraged his friends on the ship to make the most of the opportunities to explore places during port of calls.

Since 2006, he has been working as an Inventory Control Specialist for the Royal Caribbean Group, the world's second-largest cruise line operator and owner of the largest cruise ships in the world. He also managed to get side jobs while working in the ship after getting licensed as a Segway driver operator, forklift operator, and lifeboat commander-on-ship responsible for driving an evacuation boat for passengers. He then got the opportunity to travel to Antarctica in 2008 during one of the trips. He continues to be very grateful to his profession, which allowed him to visit over 100 countries and territories without spending much. He branded himself with the moniker the "Cheapatonic traveler."

Badong considers himself an ordinary person living a simple life, but nurturing big dreams. His goal is to discover more places, regardless if they are considered "touristy" or not. He is interested in meeting people cloaked with the same passion for traveling, and volunteering in various civic activities, especially when it involves going to the remotest areas in the mountains. His key interests are adventure travel, backpacking, camping, diving, hiking, street food, and traveling.

"He did not think of himself as a tourist; he was a traveler. The difference is partly one of time, he would explain. Whereas the tourist generally hurries back home at the end of a few weeks or months, the traveler, belonging no more to one place than to the next, moves slowly, over periods of years, from one part of the earth to another. Indeed, he would have found it difficult to tell, among the many places he had lived, precisely where it was he had felt most at home." – Paul Bowles, The Sheltering Sky

For him, every day is a new day and a new adventure. He encourages people to invest and discover adventure travel and exploration. As he puts it, *"Don't just travel. Explore."*

Badong's Travel Milestones:
✓ Traveled 120 out of 329 countries and territories (105 out of 193 U.N. countries)

✓ Reached 100th country and territory in 2014

✓ Visited 7th continent in 2008

✓ Completed trips to all new 7 Wonders of the World in 2019

Vanessa "Vhang" Soledad-Buechler

Qasr Kabaw, Libya

📍 **Philippine Hometown:** Tacloban City, Leyte Province

🏠 **Primary Residence:** Weggis, Switzerland

💼 **Profession:** Teacher

🗓 **Age:** 34

✈ **Traveled Continents:** 6

🌍 **Traveled Countries & Territories:** 109 (76 U.N.)

🌐 **Travel Sites:** https://www.instagram.com/vhang_dawanderer01/

https://mtp.travel/users/18801

> *"A journey is like marriage. The certain way to be wrong*
> *is to think you control it." – John Steinbeck*

Vhang was born and raised in Tacloban City, being the middle child sandwiched by two brothers. She studied elementary and secondary school at the San Jose National School, then completed a Bachelor of Science in Industrial Education from Eastern Visayas State University. She worked as a garment technologist at the Mary Our Help Technical Institute for Women in Cebu and later moved to Manila, employed as a front desk officer for the New Solanie Hotel.

As destiny would have it in 2009, she met the Swiss globetrotter Thomas Buechler, who has been to all 193 United Nations-member countries, in Manila through a friend. They started dating and traveling together to his favorite places like Sipalay, Boracay, Palawan, Sagada, Vigan, and other Philippine destinations. By 2010, at the age of twenty-four, she went with him for her first trip outside the Philippines by visiting Thailand and Cambodia. They got engaged in 2012 and got married in Switzerland in 2014.

Vhang would certainly not travel this much if it was not for her footloose husband, who motivates her a lot. They both share the same interests in photography and are both World Heritage enthusiasts. Vhang has reached a whopping four hundred and fifty-three UNESCO World Heritage sites around the world as of 2021. Her favorite activities while traveling include photography, exploring historical places, discovering the local markets, and sampling native cuisines, especially when it comes to seafood.

Vhang and her husband have a propensity to visit some of the remotest and hardest-to-reach places in the world. To reach the oldest standing mosque on the East African Coast, Kilwa Kisiwani (UNESCO) in Tanzania, for instance, took them an entire day's worth of bus ride from the former capital Dar es Salaam, where they stayed overnight in simple huts and rented a sailing boat to get to the actual site. In another instance, they traveled two months by train and public buses from Beijing via Gansu, Urumqi, Kashgar, and the rather difficult Torugart Pass to Kyrgyzstan in Central Asia. She has also traveled extensively throughout India, exploring sixteen states, including the territories in the North East and even the seven sister states (e.g., Nagaland, Assam) that were long off-limits to foreigners who required special permits. Also, she went to the remote Andaman Islands way out in the Indian Ocean, visiting Havelock and enjoying the incredible sunsets at Radhanagar beach, rated Asia`s best beach by the Time magazine in 2004.

Many more trips followed, including several in North African countries blessed with great civilizations and rich histories, such as Egypt and Morocco. Her itchy feet brought her to exotic African locations such as Zanzibar, Lamu island, Bujumbura, and Serengeti. She has been to all states in Australia, mostly traveling in old camper vans. She also traveled to New Zealand, where she has flown even to the remote Chatham island. In Europe, she has been to almost half of all regions and territories while visiting the most interesting World Heritage sites. Within eight years since her first overseas trip, she managed to reach her one-hundredth country and territory, having traveled to six continents using only a Philippine passport. These are among the reasons that she is grateful for life, often evidenced by her winsome smile. Her advice would be to keep smiling because life is a beautiful thing and there's so much to smile about. Vhang lives under the motto: *"Never get so busy making a living that you forget to make a life! We have only one life… live it!"*

Vhang's Travel Milestones:

✓ Traveled 109 out of 329 countries and territories (76 out of 193 U.N. countries)

✓ Reached 100th country and territory in 2018

✓ Visited 6th continent in 2016

✓ Visited 453 UNESCO World Heritage Sites

Ivan Henares

Trinidad, Cuba

Philippine Hometown: San Fernando, Pampanga

Primary Residence: Metro Manila, Philippines

Profession: Educator / Travel Blogger

Age: 42

Traveled Continents: 6

Traveled Countries & Territories: 108 (73 U.N.)

Travel Site: https://www.ivanhenares.com/

Ivan's family roots lie in Pampanga, though he took his elementary and secondary education at the Ateneo de Manila University. He then pursued a Bachelor of Arts in Economics, Master of Business Administration, and Diploma in Urban and Regional Planning degrees in the University of the Philippines. From a very young age, he has been a staunch advocate for the preservation of Philippine heritage. As a college student, he initiated a program that focused on the protection, conservation, and promotion of the architectural heritage of his hometown. In 2001, he was named "Most Outstanding Kapampangan for Youth Leadership" in recognition of his leadership in raising cultural awareness in Pampanga. In 2005, he received the United Nations Association of the Philippines "Outstanding Youth Leader Award." In 2006, he received "The Outstanding Fernandino Award" for Preservation of Heritage and Promotion of the Arts.

Ivan's exposure to international travel took off when he participated in the Ship for Southeast Asian Youth Program (SSEAYP) in 2002, organized by the government of Japan in partnership with the Philippine National Youth Commission. He was part of a group of youth leaders traveling on a ship sailing across the different Southeast Asian countries to promote friendship and mutual understanding.

Ivan was among the first Filipinos to take advantage of the entry of low-cost carriers in the Philippines. In April 2005, he booked a "free seat" plane ticket to Kota Kinabalu, Malaysia, and it was during this trip where he created a blog, inspired by the thought of him being a man about town for travel. As he was still doing his Masters in Business Administration at that time, blogging was more of a hobby he can do whenever he had free time. After finishing his studies close to a year later, he attended a workshop on blogging and podcasting as a political communication tool, which opened his eyes to new possibilities. From thereon, he started to create more entries and focused more of his time on using his blog as a platform for heritage advocacy. He became one of the pioneers of travel blogging in the Philippines. In 2007, his travel blog was recognized as the "Best Travel Blog" during the 1st Philippine Blog Awards. He became a rabid domestic traveler, and with his insatiable penchant for travel, he managed to visit all 81 provinces of the Philippines in 2013.

Ivan has championed the cause of Philippine heritage, helping bring heritage conservation to the mainstream by serving as one of the advocacy's most visible voices, pushing for the enactment of national laws and local legislation for heritage preservation, educating Filipinos on the importance of preserving their heritage, monitoring heritage conservation issues around the country and ensuring the protection of significant historical and cultural sites. In 2008, he was named a "Philippines 21 Fellow" by Asia Society, which selects ten promising young leaders every year to join the "Asia 21 Young Leaders Summit." In 2012, he was awarded "The Outstanding Young Men" (TOYM) in the field of heritage conservation by the President of the Philippines. In 2014, The UP Alumni Association named him one of the "UPAA Distinguished Alumni for Culture and the Arts."

Ivan's advocacy and research interests lie in heritage and tourism policy, cultural tourism in historic towns and cities, industrial heritage, and indigenous communities in the Philippines. He took on a number of roles relating to these endeavors such as becoming President of the Heritage Conservation Society (2013-2016), Trustee of Nayong Pilipino Foundation (2013-2017), Head of the National Committee on Monuments and Sites at the Philippines National Commission for Culture and the Arts (2017), Philippines national representative to The International Committee for the Conservation of the Industrial Heritage (TICCIH) since 2020, and the Secretary General of the International Council on Monuments and Sites (ICOMOS) International Cultural Tourism Committee for 2020 to 2023 after being the Vice President from 2014 to 2020.

Ivan has been active in the academia having been Assistant Professor at the Asian Institute of Tourism of the University of the Philippines – Diliman since 2016. He was also Senior Lecturer at UP AIT, University of the Philippines Open University, and the De La Salle - College of Saint Benilde. From 2017 to 2021, he studied at Purdue University in the US as a Fullbright-CHED scholar, where he completed his Ph.D. in hospitality and tourism management, and graduate certificate in environment policy. While being based in the United States, he took the opportunity to visit all the 50 states, which he completed in 2020.

Having reached his 100th country and territory in 2019, his world travels allowed him to learn best practices from other places with regard to the preservation of cultural heritage in establishing a strong national identity. This has immensely enriched his understanding of the all-important work he is set to do in expanding heritage conservation and cultural tourism advocacies in the Philippines.

Ivan's Travel Milestones:

✓ Traveled 108 out of 329 countries and territories (73 out of 193 U.N. countries)

✓ Reached 100th country and territory in 2019

✓ Visited 6th continent in 2015

✓ Completed trips to all 81 Philippine provinces in 2013

✓ Completed trips to all 50 USA states in 2020

CHAPTER 2

FILIPINO WORLD TRAVELERS USING OTHER PASSPORTS

Lourdes "Odette" Ricasa

Robinson Crusoe Island, Chile

⦿ **Philippine Hometown:** Quiapo, Manila

⌂ **Primary Residence:** Los Angeles, California

▣ **Profession:** Retired Systems Analyst

⏱ **Age:** 76

◉ **Traveled Continents:** 7

⊕ **Traveled Countries & Territories:** 295 (188 U.N.)

⊕ **Travel Sites:** https://odettericasatravels.com/

https://www.instagram.com/odettericasa/

Odette was born in Quiapo, Manila. She remembers being bitten with the "travel bug" as early as twelve when her geography teacher was describing the mountain chains of the Andes in South America, the Pyrenees of France and Spain, and the Altai in Kazakhstan. She was instantly mesmerized and told herself that she would explore all these someday. At such a young age, she was fascinated with how breathtaking the Grand Canyon, Rome, and Scandinavia looked in pictures. She once proudly declared to her dad that she would soon go to America and travel around the world.

She completed her accounting degree at the Assumption College San Lorenzo in 1965. By 1971, true to her word, she immigrated to the United States and lived in Bronxville, New York. While working in downtown Manhattan beside the Empire State Building, she took the opportunity to embark on short trips to Niagara Falls, Rhode Island, Washington D.C., and Virginia. She later moved to Los Angeles and pursued further studies in Computer Programming and Analysis at the East Los Angeles City College in 1978. During her first ten years in the U.S., her focus had been working to ensure that she made a good living and save money for her future travels. Her first "big" trip was to Russia and Israel, and since then, she became a travel addict.

She fondly recalls attending a party in Los Angeles in 1987 where she met someone working for DHL who told her that if she was willing to travel alone, she could apply for work as a courier delivery staff. When she heard that the job entailed being given at least four hours' notice to travel solo that required only a personal hand-carry bag with no checked-in luggage and flying for free, she immediately pounced on the offer. Back then, solo travel was rare, especially for women. Her first work-related trip was to Madrid. She remembers meeting the DHL representative at the LAX airport, who gave her a manifest of documents where she had to check-in sixteen huge bags. After landing in Madrid to complete the delivery, she was granted a one-week side-trip vacation in Spain by her boss.

Today, Odette is now living the good life. A modern-day renaissance woman, she's an author of six books, a successful painter, a motivational speaker, and is a prolific piano player. She has had over one-hundred and twenty paintings put on display at the Ricasa Art Gallery in San Clemente, Southern California, having apprenticed with art masters around the world from such places as Kyrgyzstan, Mauritania, Solomon Islands, Fiji, France, Vietnam, Spain, Eritrea, Cameroon, Liberia, Philippines, USA, and Saint Helena Island. Her artwork has been submitted for display in Samara, Russia, through Servas International. She plans to build an art school to give free painting lessons for children who cannot afford to pay for tuition.

As a writer, Odette's books have travel themes, and as she describes vividly, *"The marvel and sensation of a journey touches a secret satisfaction expressed in my books that will stay in my conquerable soul."* Her published works include "Unguarded Thoughts" (2002), "Excerpts from Life" (2003), "Pieces of Dreams" (2004), "Running with Echoes of Desire" (2010), "Touching the Wind" (2015), and "Love Echoes...Inspire" (2020).

Moreover, having won first prize at a Toastmasters Club International Area Division Speaking Competition in the U.S., Odette's decision to be a motivational speaker comes as no surprise. She has been invited to speak to young people across continents – from an audience of 200 students at Prince Andrew School in the South Pacific's St. Helena Island to far West in Burundi's Kamenge Youth Center to California's San Clemente High School. She loves giving back to the community, as manifested by her involvement in helping Philippine victims of Typhoon Ondoy in 2010, donating to Foyer Islamique de la Charite Orphelins in Burundi in 2013, and volunteering for sports camp service at the John Hay Elementary School in Baguio City in 2013.

Odette has her own travel club of over two hundred members based in Los Angeles, California, which meets eight-to-ten times a year for networking and sharing travel adventures. She has been interviewed by a number of mainstream media such as Time Warner's Encore TV show in the United States, Radio Antena Zaragoza in Spain, and Top Matin TV Show in the Congo. In 2019, she was awarded the prestigious "Outstanding Artist Traveler" during the Entrepreneurial and Cultural Heritage Awards by the Sinag Lahi USA organization. Odette resides in Los Angeles, California. She has three children, three grandchildren, and three siblings. She also speaks three languages – English, Filipino, and Spanish. She's been traveling to Spain twice a year since 1987 up until the pandemic. The longest duration of her world travel has been seven weeks, traveling solo most of the time.

> *"Only those who can see the invisible can achieve the impossible." – Albert Einstein*

The typical question she gets asked a lot is, *"Are you not afraid to travel alone?"* Her response has always been "no." She always maintained that fear is not real; it exists only in our thoughts of the future. She constantly prays to God for guidance and wisdom in her daily life and her world travels. Before the pandemic happened, Odette was scheduled to travel to Libya, Sudan, Chad, Central African Republic, and Iraq, which would make her the first Filipino to travel in all one hundred and ninety-three United Nations countries.

Her philosophy in life is to simply ask questions and not seek answers. After all, the world is her classroom. She is passionate about seeing more of it in the coming years. And if one would ever feel fear and doubt about going out into the world, Odette's words of wisdom would always be: *"Yes, you can do it!"*

Odette's Travel Milestones:

✓ Most Well-Traveled Filipino

✓ Traveled 295 out of 329 countries and territories (188 out of 193 U.N. countries)

✓ Reached 100th country and territory in 2002, 150th in 2005, 200th in 2011, and 250th in 2013

✓ Visited 7th continent in 2015

✓ Completed trips to all UN countries in North America in 2011, Europe in 2012, South America in 2014, Oceania & Pacific in 2020

✓ Completed trips to all new 7 Wonders of the World in 2014

Luisa Yu

Badlands National Park, South Dakota, USA

📍 **Philippine Hometown:** Tacloban City, Leyte

🏠 **Primary Residence:** Miami, Florida, USA

💼 **Profession:** Retired Medical Technologist, Travel Advisor

📅 **Age:** 76

✈ **Traveled Continents:** 7

🌐 **Traveled Countries & Territories:** 223 (161 U.N.)

🌐 **Travel Site:** https://www.instagram.com/luisa_yu14/

Luisa was born in Dulag, Leyte, to a middle-class, Chinese-Filipino family. After completing her Bachelor's degree in Science from the University of Santo Tomas, she migrated to the United States in 1967 through an exchange student program in St. Louis, Missouri, where she graduated with a Bachelor of Science degree in Medical Technology. Subsequently, she was granted a license to practice as a certified supervisor in Laboratory Medical Technology by the American Association of Clinical Pathology. As she could not go outside the United States because of her immigrant restrictions at the time, she started traveling via Greyhound bus services and visited most of the states as a backpacker. By 2020, she was able to visit 45 out of 50 states in the United States.

After her formal schooling, she moved to Miami and spent the next thirty years working as a healthcare professional in various roles (i.e., laboratory night supervisor, blood bank administrator, operating staff, etc.) for such renowned health institutions as the Jackson Memorial Hospital, University of Miami, Hialeah Hospital, and Mt. Sinai Hospital Medical Center. She later received her real estate license in the state of Florida and became a practicing realtor on the side.

Always been financially self-supporting since she came to the United States, she considers herself fortunate that her jobs allowed her to travel to other parts of the world and become a freelance travel consultant. For eleven years, she worked as a travel consultant at All Ways Travel in Miami and even set up Florida International Travel, Inc. as an owner and manager before she sold the business. Since 1996, she has been working for Westchester Travel and has had the opportunity to attend several international meetings, trade shows, conferences, workshops in different countries worldwide. She had the unique opportunity of traveling with the late Klaus Billep, former Travelers' Century Club Chairman, who personally extended the invitation for her to join the club as a member. In 1982, she successfully circumnavigated the world for three months and became an official member of the Circumnavigators Club in 2014.

Luisa is a member of a number of renowned global travel organizations, such as the American Society of Travel Advisors (ASTA), Cruise Lines International Association (CLIA), and Pacific Asia Travel Association (PATA). She also served as a member of the Miami International Folk Festival Board of Directors from 1978 to 1981. For the last 20 years, Luisa has been a member of the Travel Industry Association of Florida, having assumed the role of Vice-President for seven consecutive terms and, later, President for one term. In appreciation of her nineteen-year service as Programs Chairperson of the association from 1999 to 2018, Luisa was honored with a complimentary lifetime membership.

Her main interests include travel, photography, music, dancing, and volunteer work. For her, to live a life well-lived is to explore the world and experience humanity in all its aspects. She earnestly believes that traveling is discovering what real life is all about. A fascinating journey, therefore, makes one a happier person.

> *"Remembering that I'll be dead soon is the most important tool I've ever encountered to help me make the big choices in life. Almost everything — all external expectations, all pride, all fear of embarrassment or failure — these things just fall away in the face of death, leaving only what is truly important. Remembering that you are going to die is the best way I know to avoid the trap of thinking you have something to lose. You are already naked. There is no reason not to follow your heart."* —Steve Jobs

Luisa is a clear manifestation that age is never a hindrance to being a world traveler. Despite being a septuagenarian, she has demonstrated extraordinary vim, vigor, and vivacity that outshine contemporaries half her age.

She is, indeed, a hodophile personified.

Luisa's Travel Milestones:
- ✓ Traveled 223 out of 329 countries and territories (161 out of 193 U.N. countries)
- ✓ Reached 100th country and territory in 1996, 150th in 2012, and 200th in 2019
- ✓ Visited 7th continent in 2010
- ✓ Circumnavigated around the world in 1982
- ✓ Completed trips to all new 7 Wonders of the World in 1990

Sonriza "Riza" Rasco

Mundari tribe, Terekeka County, South Sudan

⊙ **Philippine Hometown:** Los Baños, Laguna

⌂ **Primary Residence:** Bay, Laguna

▣ **Profession:** Bioengineer, Intellectual Property Management Professional and Social Entrepreneur

⊙ **Age:** 50

⊘ **Traveled Continents:** 6

⊛ **Traveled Countries & Territories:** 224 (173 U.N.)

⊕ **Travel Sites:** https://www.instagram.com/rizarasco

https://exploreafrica4impact.com/

Riza was born in Los Baños, Laguna, to parents who were university employees and scientists. Her father is a plant breeder, her mother was a nutritionist, and her brother is a chemical engineer. It doesn't come as a surprise that she pursued a similar education path — finishing a Bachelor of Science in Agriculture (with honors) as well as a Master of Science in Agriculture Biotechnology from the University of the Philippines in Los Baños.

Having grown up in Los Baños, Riza recalls that as a kid, she was prohibited from going to the forest and climbing trees and playing on a busy rail track with street kids, but she went anyway. She always wanted to try something new and to go on an adventure. Riza's first plane ride and trip abroad was to Ithaca, New York, when she was just five years old. She left the Philippines in 1993 (that was her first trip abroad as an adult) at age twenty-two. She pursued her Ph.D. in Cell and Molecular Biology/Bioengineering from The University of Nottingham in the United Kingdom as a Rothamsted Research scholar. She also completed a Certified Executive Program for Global Business Management from the prestigious Wharton School in the University of Pennsylvania, as well as a Certified Program on Negotiation at Harvard Law School. She ventured a career in biotechnology, working as a research scientist in both the public and private sector for the Rothamsted Research institute and the DuPont Company in the United Kingdom and the United States, respectively. She later worked as a manager and senior associate in intellectual property licensing and commercialization at the DuPont Company and John Hopkins Technology Ventures, respectively.

When her mother passed away in 2013 after losing a battle to breast cancer, she realized that life is short so she committed to travel more; she began a blog called Midlife Funk because, at the time, it reflected her depressed and restless mood. Her mother's death gave her the strength and courage to quit her seventeen-year stint as a corporate employee to pursue her dreams and travel the world. After taking a personal sabbatical leave from her professional career as a biotech executive from November 2015 to October 2016, she embarked on expeditionary tours to Africa, traversing the continent and visiting approximately thirty countries.

In 2018, she founded Explore Africa for Impact, a tour operator and social venture company based in West Africa that aims to spur more foreign travel to the African continent for the benefit of the local communities there, particularly in the areas of education and women empowerment. As she puts it, "My heart is in the Philippines, but my soul is in Africa." Ethiopia is her top country though she has many, many favorite countries. With an open mind and heart, she believes every country and place on earth has something interesting to offer to visitors, whether it be about the people, the landscape, the flora and fauna, the food, the stories that can be enjoyed. Possessing three passports — Philippines, United States, and the United Kingdom — she frequently travels to these three countries where she enjoys a home base.

After twenty-six years of working and traveling overseas, she came back to the Philippines to make it her main home base since 2019, working as an independent consultant and managing director of her enterprise specializing in intellectual property and technology transfers, as well as licensing and commercialization.

Riza loves doing volunteer work and helping communities. She hiked up the Mount Everest base camp in Nepal to raise funds for the Lymelight Foundation. Through her company, Explore Africa for Impact, she trained African women to become tour guides and initiated building a school in the Bondorfulluhun Village at the Kono district in Sierra Leone. In line with her belief that the privilege to travel the world comes with the duty to give back to one's country, she co-founded the Philippine Global Explorers and became its Founding Chairman and Charter President.

Riza describes herself as a "multipotentialite," a person who has many different interests and creative pursuits. Apart from travel, her other interests lie in photography, cooking, yoga, running, and dance. One of her travel photos was used as a poster ad by the Ministry of Tourism in Panama. A fun travel tidbit about Riza is that she collects photos of her being on top of military tanks.

Having served as a former Zumba instructor, she even trained with its founder, Beto Perez. Riza is a believer in the "Work Hard, Play Hard" philosophy. Traveling the world requires a significant

"We are travelers on a cosmic journey, stardust, swirling and dancing in the eddies and whirlpools of infinity. Life is eternal. We have stopped for a moment to encounter each other, to meet, to love, to share. This is a precious moment. It is a little parenthesis in eternity." – Paulo Coelho, The Alchemist

amount of money. Since she was not born with a silver spoon, she had to study hard, build a professional career, move abroad, work harder and save her earnings for many years before she could travel the way she does now. There is no magic bullet to funding travel, as it can only be achieved through hard work and prudent management of resources.

Riza's Travel Milestones:

✓ Traveled 224 out of 329 countries and territories (173 out of 193 U.N. countries)

✓ Reached 100th country and territory in 2016, 150th in 2019, and 200th in 2020

✓ Visited 6th continent in 2006

✓ Completed trips to all UN countries in Europe and Africa in 2021

April Peregrino

Burning Man, Black Rock City, Nevada, USA

📍 **Philippine Hometown:** Dulag, Leyte

🏠 **Primary Residence:** Chicago, Illinois, USA

💼 **Profession:** Pharmacist

📅 **Age:** 42

✈ **Traveled Continents:** 7

🌐 **Traveled Countries & Territories:** 185 (159 U.N.)

🌐 **Travel Site:** https://www.instagram.com/everyotherweek/

> *"You see things; and you say, 'Why?' But I dream things that*
> *never were; and I say, 'Why not?'" – George Bernard Shaw*

Born in Chicago to Filipino parents who immigrated from the Philippines to the United States in the 1970s, April inherited her father's wanderlust genes, apart from his surname, which means "pilgrim" in Spanish. Her father is a ship captain from Leyte, and her mother was a nurse from Samar. When she was six years old, her parents decided that the first time she would ride an airplane was through a trip to the Philippines. When she was in sixth grade, she was surprised to learn from a friend that she didn't need a passport to fly to another U.S. state.

April went on to study at Trinity High School, a private all-girls Dominican College Preparatory School nestled within the suburbs of River Forest, Illinois. Under this rigorous environment of self-direction toward responsible participation, her core beliefs of self-empowerment and discipline were forged. April never takes "no" for an answer: she always finds a way around an obstacle or waits until she can eventually get what she wants to fulfill a goal.

At the age of twenty one, one of her best friends fundamentally transformed the way she saw the world when he introduced her to Europe, with France being the first country she visited outside of the Philippines and the United States. At the age of twenty-three, April and another friend decided to go to Peru to embark on the iconic four-day, three-night Inca Trail hike; that trip shaped her travel style, preferring to add more layers to her travel experience for a richer cultural experience every step of the way. Four months later, during the month between her third and final year of pharmacy school, she took her first solo trip to Brazil which became an empowering experience. It taught her to live on the edge and explore destinations, using only her wit and intelligence, as well as her intuition, as she immersed herself in a country which was supposedly "dangerous" where she didn't speak the language. It allowed her to feel more comfortable in her own skin and more self-confident.

After earning her Doctorate of Pharmacy from the University of Illinois in Chicago, she began her career as a pharmacist. Since 2005, she's been working as an overnight retail pharmacist where she's given a full week off after working for seven consecutive nights. This schedule paved the way for her to embrace a bi-weekly nomadic lifestyle.

Her mother's unexpected passing in October 2009 pivoted her perspective and priorities, preferring to live her life to the fullest every day. She came to acknowledge she only has one life to live, and she realized her life's purpose. She tries her best to fulfill it in her lifetime. Acting on this conviction, she made it a personal goal to travel to all the seven continents in one calendar year, which she achieved in 2013, arguably making her the first Filipina to do so. She finally reached her one-hundredth country and territory in 2016. While attending her second Burning Man event in 2018, she got her epiphany to travel to all one-hundred and ninety-three United Nations countries. A few of her travel highlights included traveling via overland truck and tent-camping from Eastern Africa to Southern Africa for four months, embarking on six high-altitude treks over four thousand meters above sea level, including Mount Kilimanjaro, Mount Kinabalu, and the Everest Base Camp. She has even organized a tour for four Americans to go to Mogadishu, Somalia, where she accompanied them. For this compulsive sojourner, a new destination means meeting more like-minded people, increasing her traveler tribe, and making her bucket list longer.

Other than travel, April loves to run and hike. She ran one full marathon in San Diego, U.S.A., and several half-marathons, including one in Pyongyang, North Korea. She also practices yoga, devours books, and has an affection for learning new languages. Her greatest fascination involves meeting people and the psychology behind their decisions. She loves to share her empowerment story and spread positivity which both align with the three philosophies that she embraces, which are to: "Go big or go home;" "Love yourself;" and "Live life to the fullest. You only have one life!"

April's Travel Milestones:
- ✓ Traveled 185 out of 329 countries and territories (159 out of 193 U.N. countries)
- ✓ Reached 100th country and territory in 2016, and 150th in 2019
- ✓ Visited 7th continent in 2013
- ✓ Completed trips to all new 7 Wonders of the World in 2016

 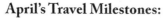

Raoul Bernal

Nine Arch Bridge, Ella, Sri Lanka

📍 **Philippine Hometown:** Cebu City

🏠 **Primary Residence:** San Francisco, USA

💼 **Profession:** statistical programmer
and biostatistics consultant

🧭 **Traveled Continents:** 6

🌐 **Traveled Countries & Territories:** 162 (121 U.N.)

🌐 **Travel Site:** https://www.instagram.com/raoulbernal/

> *"To get away from one's working environment is, in a sense, to get away from one's self; and this is often the chief advantage of travel and change." – Charles Horton Cooley*

Born in Quezon City, Philippines, Raoul was raised in Cebu, where he completed grade school and high school studies. He pursued statistics and computer science at the University of the Philippines in Los Baños, where he eventually taught as a junior instructor. He got selected as a delegate to the Philippines-Japan Friendship Programme in Tokyo, which was his first international travel. Shortly after, Raoul moved to Illinois, USA, to pursue his graduate studies in applied statistics. After completing his degree, he worked as a QA Analyst for SPSS, a statistics software company in Chicago. He traveled to Europe (mainly the United Kingdom, Italy, and France) for the first time after obtaining his permanent resident status in the United States.

Raoul relocated to California in the late 1990s to pursue biostatistics coursework at the University of California in Los Angeles (UCLA). He worked as a statistical programmer analyst for several large biotech and pharmaceutical companies like Gilead, Amgen, Genentech/Roche, and AstraZeneca. He worked his way from being a project lead to manager/director before he quit working full-time. He considers himself fortunate to have accrued plenty of vacation days at work, which he used to travel extensively. During his visits to the Philippines, he also took side trips across Southeast Asian countries.

In 2000, he became a lifetime member of Mensa, the largest and oldest high-IQ society in the world. He attended international Mensa meetings and regional gatherings, which is how he expanded his travel coverage and adventures.

Raoul fondly remembers that during a work offsite meeting in early 2017, they were tasked to come up with a photograph or selfie along with a description for an improvement plan or self-promise. He wrote: *"I promise to start living the rest of my dreams, however far wanderlust takes me."* The photographer had him pose as a traveler pretending to read a map, and photoshopped a countryside scene with a winding railroad in the background. It truly depicted what he felt he wanted to pursue. It seemed odd to his colleagues as his "self-improvement plan" was not work-related, but they always knew about Raoul being passionate for travel and adventure. Little did they know that he would resign soon after that offsite meeting to embark on a "self-imposed travel sabbatical."

About a year later, he found himself visiting the Nine Arches Bridge in Ella, Sri Lanka. As his tour guide led him along the winding railroad track, a sudden *déjà vu* feeling hit him – it was almost as if he was reliving the same setting the photographer created for him a year earlier. It was very surreal, but it was also an overwhelming feeling of accomplishment, as it validated his decision to quit full-time work to pursue his travel plans. In a span of twenty months, he covered over fifty countries, thirty of which were first-time visits. As a result, he qualified for the Travelers' Century Club in 2017. He subsequently made it to the silver status having visited one hundred and fifty countries and territories with his visit to the Easter Island in January 2020.

"We wanderers, ever seeking the lonelier way, begin no day where we have ended another day; and no sunrise finds us where sunset left us. Even while the earth sleeps we travel. We are the seeds of the tenacious plant, and it is in our ripeness and our fullness of heart that we are given to the wind and are scattered." - Kahlil Gibran

Raoul loves visiting museums in major cities, specifically those that showcase the history and culture of the country he is visiting. For him, it is the best way to learn about the country's past and its aspirations for the future. He is an avid collector of various mementos, ranging from coins and currency bills (including the European currencies before the Euro was introduced), Hard Rock Café shot glasses, pins, and classic t-shirt collections. He is also a member of an international postcard group that collects and trades postcards as souvenirs.

Raoul enjoys the challenge and excitement of traveling to new places, as well as returning to places to meet up with friends and other kindred spirits who share his travel avocation. He believes one needs to define their passions in life to keep one motivated. His life's philosophy is to amass and enjoy travel experiences more than material things, challenge oneself out of one's comfort zone, and live life to the fullest with the opportunities one is given. *"Life is like a game of poker; you can only play the hand that you're dealt."*

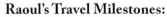

Raoul's Travel Milestones:

✓ Traveled 162 out of 329 countries and territories (121 out of 193 U.N. countries)

✓ Reached 100th country and territory in 2017, and 150th in 2020

✓ Visited 6th continent in 2005

✓ Completed trips to all UN countries in Europe in 2018 and South America in 2020

Brian Banta

'Gates of Hell' Darvaza Gas Crater, Turkmenistan

📍 **Philippine Hometown:** Muntinlupa City, Manila

🏠 **Primary Residence:** Metro Manila, Philippines

📷 **Profession:** Entrepreneur / Restaurateur

🕐 **Age:** 42

✈ **Traveled Continents:** 7

🗺 **Traveled Countries & Territories:** 142 (117 U.N.)

🌐 **Travel Site:** https://www.instagram.com/travelingbants/

Brian is the youngest of four brothers in a modern-day, blended family. A gifted student, he went to the International School Manila before pursuing a double-degree in Economics and International Relations at Clark University (Worcester, Massachusetts) in the United States, graduating with *summa cum laude* honors in 2001. He is a member of Phi Beta Kappa, the oldest academic honor society in the United States, and Omicron Delta Epsilon, an international honor society in the field of economics. He later received his Master's degree in International Trade and Investment Policy from the George Washington University (Washington, D.C.) and completed an entrepreneurship certification at the Asian Institute of Management in Manila, Philippines.

After completing his studies, Brian immediately worked as a project director for multiple international publishing companies which involved leading, researching, interviewing, and publishing business country reports in six out of the seven continents. These reports were published in some of the world's most respected media outlets, including *Foreign Affairs* (USA), *The Japan Times* (Japan), and the *Forbes* Group, and have allowed him to work and live in over twenty countries.

Brian is a genuine entrepreneur at heart; he has found fulfillment in setting up restaurants, bars, and coffee shops throughout Manila, which serve a wide demographical range. These include the 121 Restaurant chain, Salamangka, and LIT Manila, a Japanese whisky bar that serves the largest selection of Japanese spirits in the Philippines. Always interested in new ideas, Brian has also dabbled in e-commerce and the importation of foreign foods.

His variety of interests cover world cuisine, specialty coffee, Hollywood movies, NBA basketball, comic books, and other popular geek trivia. As a matter of fact, he used to host a weekly quiz night at 121 that ran for twelve years! He has also dabbled in hosting gigs, has emceed multiple weddings, and is a known mic-hog at karaoke and "band-eoke" nights.

A devoted husband and father to two bustling boys, Brian's other love is adventuring through travel. He has been fortunate enought to live to travel and travel to live. He developed his passion from traveling with his parents, who encouraged him to visit as many countries as he could; in fact, he used to compete with his dad to see on who could visit more postage stamp-issuing countries. Brian won several years back when he overtook his father's also commendable 87 countries.

Learning to dive in Australia's Great Barrier Reef is one of his most impactful travel experiences which led to an infatuation with scuba diving in his twenties. He has always had a great love for animals and unendingly searches for new encounters when traveling, such as swimming with pink dolphins in the Amazon, going on jaguar-spotting safaris in the Pantanal (Brazil), petting tigers in Thailand, witnessing giant turtles lay eggs in Oman, fishing with penguins in the Galapagos islands in Ecuador, chasing humpback and minke whales in Antarctic waters, and observing Komodo dragons in the wild.

Some of Brian's most memorable world experiences include traversing the Central Asian Silk Road, climbing the Uluru inselberg in Australia and Mt. Fuji in Japan, off-road driving to visit moais in Easter island, exploring the driest place on Earth in the Atacama Desert of Chile, climbing the Mayan ruins of Tikal in Guatemala, kayaking the alpine blue Soča river in Slovenia, and hopping from ambrosian street tacos to Teotihuacan pyramids, to the Basilica of Our Lady of Guadalupe all in the same day in Mexico City. A few of his unforgettable trips to Antarctica and the Camino de Santiago pilgrimage in Spain were even chronicled in travel articles published on ANCX.

Brian's Travel Milestones:

- ✓ Traveled 142 out of 329 countries and territories (117 out of 193 U.N. countries)
- ✓ Reached 100th country and territory in 2018
- ✓ Visited 7th continent in 2020
- ✓ Completed trips to all new 7 Wonders of the World in 2014

Donalito "Dondon" Bales, Jr.

Cayos Zapatilla, Bocas del Toro, Panama

📍 **Philippine Hometown:** Tacloban City, Leyte

🏠 **Primary Residence:** Melbourne, Australia

📷 **Profession:** I.T. Project Manager

😎 **Age:** 43

🧭 **Traveled Continents:** 7

🗺 **Traveled Countries & Territories:** 141 (107 U.N.)

🌐 **Travel Site:** https://www.instagram.com/explorenextlevel
https://www.explorenextlevel.com

Born in Tacloban City, Dondon is the eldest of four siblings. When he was growing up, his father worked as a local tour guide driver while his mother was a housewife. As a kid, he wanted to emulate Jose Rizal the Philippine national hero, who could speak many languages and has traveled quite extensively during his time. His grandmother gifted him with a collection of stamps and coins from different countries, which started his hobbies in philately and numismatics.

Even during his elementary years at the Leyte Normal University (formerly Leyte State College), Dondon was fascinated by other countries that he made a research paper about the Pyrenees mountains in France. His student-teachers predicted that he would be an "ambassador" one day. A gifted student, he competed and won in the prestigious National QuizBee Science Category and represented the Philippines during the first-ever World Quizbee hosted in Manila, where he landed 2nd place. His impressive win opened greater academic doors for him as he was admitted as a national scholar of the Philippine Science High School in Diliman, Quezon City. There, he discovered his love for computers, and by his senior year in high school, he was part of a team that successfully competed in a national computer programming competition to represent the Philippines in 1993 at a global coding contest in Hongkong at the age of 16. That was his first overseas trip.

Because of his sterling academic credentials, Dondon was given a full college scholarship by the Ayala Foundation to pursue a Bachelor's Degree in Computer Science in the Ateneo de Manila University.

And despite not having a computer of his own, he still graduated with academic honors, finishing with an "Honorable Mention" distinction. Dondon also won first place in a national essay writing competition about migration hosted by the Department of Foreign Affairs, which was an auspicious portent of things to come.

In 1998, when he was just 20 years old, Dondon was offered a job by Accenture (formerly Andersen Consulting). During his first month at work, he was already sent to St. Charles, an hour west of Chicago, Illinois, USA, for a month-long training as a consulting analyst. This was his first long-distance trip, and he truly had a "We are the World" moment when he was asked to pose for a "class photo" with his work colleagues raising their respective flags of origin.

As his career was thriving in Accenture, he also got involved in Junior Chamber International (JCI), a global, non-profit organization for young people to develop leadership skills, meet inspiring individuals and make a real impact in local communities. He volunteered and took on a number of leadership roles in community projects and training programs. He was conferred with a lifetime membership as a JCI Senator in 2005. Winning the "Most Outstanding JCI Member of the Philippines" in 2006 as part of JCI Makati chapter, he represented the country and won the "Most Outstanding JCI Member for the Asia-Pacific" in Takamatsu, Japan; he was also selected as the "World Congress First Timer Award" in Seoul, South Korea in 2006. It was in his stint in JCI that Dondon began to take advantage of the organization's international events to satisfy his desire for travel while at the same time, expand his network around the world.

Through his work in Information Technology, he was given the opportunity to be assigned to Japan, Italy, USA, and Australia. It was at this time that he was enamored by the values and quality of life in Australia; as such, he took a leap of faith and migrated alone to Australia and settled in Melbourne in 2007, where he worked for Accenture Australia. Two years later, he moved to Sydney for a work project while continuing his career in JCI, even being elected as JCI Sydney Chapter President in 2011 as well as winning the "Most Outstanding President" and "Most Outstanding Trainer" awards. He was later elected to the JCI Australia National Board and was also appointed as International Officer tasked with liaising with other chapters worldwide.

Dondon decided to take a 21-month sabbatical from his work which began in late 2011. He used this time to travel around seventy-five countries and territories, crisscrossing seven continents, circumnavigating around the world. When he came back to work in late 2013, he got assigned overseas for over a year living in the US, UK, China, Philippines, Indonesia, and Papua New Guinea, where he got to circumnavigate the world a second time. Working remotely from several countries, he had put all his belongings in storage and traveled with no permanent address since 2012 until he settled back in Sydney in mid-2015. By then, he had already reached his one-hundredth country and became the coordinator for a potential Australia chapter for Travelers' Century Club. He co-founded the Philippine Global Explorers travel community and is a Founding Director and the

Charter Vice President of Operations, and was responsible for the creation of the organization's website and the Zoom call sessions.

He continued to work as a Senior Manager in Accenture Australia until 2019 and won one of the best project awards in the company in 2017. In 2020, he completed a management course at Harvard Business School and worked as a Project Manager for Seven Consulting, Australia's best program delivery company. He also founded his travel social enterprise start-up, Explore Next Level, as part of his vision to be a philanthropist. Dondon also signed up with Virgin Galactic in 2020 to become a spacefarer, eagerly awaiting his turn to go to space within the decade.

Dondon has the profile of an unlikely world traveler: from someone who came from very humble beginnings in a Philippine countryside; financed his education purely through scholarship grants; became his family's breadwinner as a young adult; worked as a global corporate executive; migrating alone overseas; possessed a phobia for riding turbulent planes and stray dogs; never knew how to swim properly; finished last in his 1st full marathon in 2017; not possessing a full driver's license until the age of 42; never knew how to use a DSLR camera; or not knowing how to cook homemade dishes until the pandemic lockdown.

Living life at a fast pace, he stayed committed to his dream of being a world traveler, doing whatever it took to get there despite the enormous odds. Describing himself as an "experience connoisseur," he celebrates life in constant search of meaning, hoping to become a testament to ordinary people that living an extraordinary life is always possible, only if one chooses to do so.

"Oh, the joys of travel! To feel the excitement of sudden departure, not always knowing whither. Surely you and I are in agreement about that. How often did my life seem concentrated in that single moment of departure. To travel far, far—and that first morning's awakening under a new sky! And to find oneself in it—no, to discover more of oneself there. To experience there, too, where one has never been before, one's own continuity of being and, at the same time, to feel that something in your heart, somehow indigenous to this new land, is coming to life from the moment of your arrival. You feel your blood infused with some new intelligence, wondrously nourished by things you had no way of knowing." – Rainer Maria Rilke

Dondon's Travel Milestones:

✓ Traveled 141 out of 329 countries and territories (107 out of 193 U.N. countries)

✓ Reached 100th country and territory in 2015

✓ Visited 7th continent in 2012

✓ Completed trips to all UN countries in Europe and Southeast Asia in 2019

✓ Circumnavigated around the world in December 2011 to January 2013, and in March 2014 to August 2014

✓ Completed trips to all new 7 Wonders of the World in 2013

Hegina "Henna" Abueva-Fuller

Pashupatinath Temple, Kathmandu, Nepal

📍 **Philippine Hometown:** Guiuan, Eastern Samar and Tacloban City, Leyte

🏠 **Primary Residence:** Kansas City, Missouri, USA

📷 **Profession:** Business owner

📅 **Age:** 51

✈ **Traveled Continents:** 6

🗺 **Traveled Countries & Territories:** 143 (103 U.N.)

🌐 **Travel Site:** https://www.culturalwanderer.com

> *"Little do ye know your own blessedness; for to travel hopefully is a better thing than to arrive, and the true success is to labour." – Robert Louis Stevenson*

Henna is a true-blooded Waray: her mom comes from Tacloban City, and her dad is from Guiuan, Eastern Samar, both situated in the Eastern Visayas region whose residents belong to a large Austronesian family. While close-knit, her extended family is geographically spread out across the National Capital Region and Mindanao. Her dad's siblings, for instance, studied, lived, worked in Butuan City and Manila. Her mom's only sister also lived in Manila. When she was a young child, Henna's grandparents would almost always bring her and her siblings to visit these places, especially during summertime. Henna was introduced to 'seafaring' very early, sometimes less memorable: she would often get bouts of motion sickness and would vomit on the road. These early travels were foreshadowing the trajectory of her world trips over the next decades.

Henna studied at Guiuan Elementary School, moved to Agusan National High School in Butuan City but eventually completed her secondary schooling at the Assumption Academy of Samar in Guiuan, always ending up being a student leader and on the honor roll being consistently at the top 5. Aside from pursuing academic excellence, Henna was into arts, ballet, piano, drama, and theater. She later received her Bachelor's degree in Secondary Education at the Leyte Normal University. She was already on her way to secure her Master's degree in Public Administration at the University of the Philippines – Tacloban but was cut short when she seized the opportunity to migrate to the United States with a Philippine passport. She was working as a Marketing Specialist at Pag-IBIG Fund / HDMF with limited options being a contractual employee, so she took the opportunity of migrating to the USA.

This happened through an auspicious hiking trip to El Nido, Palawan, in the Year 2000. Henna met Steve Fuller, a city judge, and renowned lawyer, and a world traveler who has already visited over 100 United Nations member countries, and he fell for her instantly. They married in 2002, and she moved with him to Kansas City, Missouri, USA. As a housewife, she would assist at her husband's law office on various activities and dabbled in a modest real estate and rental business where she learned how to become an entrepreneur. This exposure led her to meet a lot of successful people and developed connections and relationships, and of course, inspiration. Henna is also engaged in different charitable organizations, volunteer groups, and travel clubs.

Her husband, having run 240 marathons, introduced her to yoga and running. While not a full-fledged marathoner, she still runs in the 5k and 10k categories on occasion. A week after she landed in the United States, she enrolled in an Ashtanga Yoga training course in Lake Tahoe, California. Today, she leads yoga and wellness trips, leads walking/hiking/forest bathing events but with a yoga element into it. Through her husband's encouragement, Henna embarks on tours at least three-to-four times a year, each ranging from two-to-three weeks. In 2014, she had the unique opportunity to circumnavigate the world. In 2016, she reached her one-hundredth country in Uruguay. For her, the highlight of every trip is falling in love with the place and its people and going back over and over to the same place and never getting tired of it. She particularly loves Portugal, Spain, France, Mexico, USA, the Philippines and had the privilege of visiting its regions and provinces.

She later decided to start her own travel company, the Cultural Wanderer. Using her yoga fitness studio (NKCYOGA) as a starting point to reach out and begin yoga retreats overseas, she established her Cultural Wanderer company to offer unique cultural experiences and adventures to travelers seeking to explore the world. For four years already, she takes her guests to different countries and introduced new travelers from the Midwest to see the world through her experiences. She even started a Facebook Group, which she christened "Kansas City Women Travel" a few weeks before the pandemic began.

Henna considers all her business endeavors as a way of giving back to the community. Travel makes her proud and makes her humble at the same time. She likens herself to a little pebble in a vast sea capable of making bigger waves for change.

Henna's Travel Milestones:

- ✓ Traveled 143 out of 329 countries and territories (103 out of 193 U.N. countries)
- ✓ Reached 100th country and territory in 2016
- ✓ Visited 6th continent in 2008
- ✓ Circumnavigated around the world in 2014

Kit Reyes

Shwedagon Pagoda, Yangon, Myanmar

📍 **Philippine Hometown:** Metro Manila

🏠 **Primary Residence:** Toronto, Canada

📷 **Profession:** Producer/Filmmaker

🕑 **Age:** 44

✈ **Traveled Continents:** 5

🗺 **Traveled Countries & Territories:** 132 (117 U.N.)

🌐 **Travel Site:** https://www.instagram.com/shutterkit/

Kit was born in the Philippines and moved to Canada with his entire family in 1993. As a child, he would always have a fixation on travel atlases and maps. By seven, he could already name every capital city of every country. That same year, he took his first flight to visit his grandparents in the USA; that memorable trip was to define the rest of his life.

He has been in the honor class throughout his stay, from elementary to high school at La Salle Greenhills, San Juan City. An acclaimed student prodigy in his school, he was a perennial fixture in the Student Council and was part of a pull-out program for the academically gifted. When he moved to Canada, he entered Ryerson University and graduated with a degree in Radio and Television Arts. Rogers Broadcasting conferred him the prestigious John Webb Q.C. Rogers Multicultural Scholarship, an entrance award bestowed on Canadian minorities who have made a significant difference in their communities.

Kit later worked in the entertainment industry, notably as an intern at the "Late Show with David Letterman" in New York City and a number of Canadian television programs based in Toronto. He briefly dabbled in the corporate world but realized it wasn't for him. In 2001, he left for Europe and worked as a media and television consultant and a government relations manager for a company based in Belgium for over eleven years.

As an experienced film and video producer, he managed complex productions all over the world and has worked in a variety of roles, including producer, writer, director, project manager, and script editor, for programs primarily geared towards the international market. As a digital sales professional, he worked closely with many of the world's biggest advertisers and agencies in developing video solutions that build strong businesses and brands. He has interviewed hundreds of heads-of-state, government officials, and CEOs of Fortune 500 companies for various media networks such as CNBC, Bloomberg, Fox, Channel News Asia, and Discovery Channel.

Apart from being a respected global producer and filmmaker, Kit is also a man of many talents. He is a musician gifted with a perfect pitch (the ability to identify or re-create a given musical note without the benefit of a reference tone) and used this gift in performing with various Christian bands all over the world, including the Office of Catholic Youth for the Archdiocese of Toronto. He is also a talented photographer. As he proudly puts it, "If I've been there, I've shot it."

Kit considers being able to combine his passion for travel and his career as his key personal accomplishments. He loves to travel spontaneously; in fact, he only packs on the day of his flight, never collects souvenirs from any of his travels, and is the type to throw caution to the wind by jumping at the next travel opportunity without the need for preparation.

While quarantined at home during the COVID-19 pandemic, Kit re-pivoted his travel objectives: no longer does he aspire to complete all 193 United Nations member-countries; rather, just focus on traveling to places that he really loves to be in, even if that means going back to the same places several times.

"There is no happiness for him who does not travel, Rohita! Thus we have heard. Living in the society of men, the best man becomes a sinner... therefore, wander!... The fortune of him who is sitting, sits; it rises when he rises; it sleeps when he sleeps; it moves when he moves. Therefore, wander!" – Aitareya Brahmanan, Rigveda

When asked of his life's philosophy, he offered two words: "Never Settle."

Kit's Travel Milestones:
- ✓ Traveled 132 out of 329 countries and territories (117 out of 193 U.N. countries)
- ✓ Reached 100th country and territory in 2012
- ✓ Visited 5th continent in 2002

Jazmin "Jazz" Gaite

Tiki Tuhiva, Tiaohae, Nuku Hiva island, Marquesas, French Polynesia

📍 **Philippine Hometown:** Manila

🏠 **Primary Residence:** Tarzana, California, USA

💼 **Profession:** Retirement home business owner

📅 **Age:** 69

✈️ **Traveled Continents:** 6

🗺️ **Traveled Countries & Territories:** 128 (90 U.N.)

🌐 **Travel Sites:** https://www.facebook.com/jazmin.gaite/

Jazz is the eldest child of nine children and three half-siblings of lawyer-parents. Her love for travel began when she found her mother's stamp collection. Not only did it contain first-day issues of Philippine stamps but also philately from places that she has never heard of, such as Magyar Posta, Sverige, and Andorre. As a child, she wondered how those places looked like and what the people in those countries were like. When she was about seven, her father would bring home boxes full of Philippine one-centavo coins for them to count and asked her to gather them in fifty-centavo stacks for him to deposit in the bank. Mixed with these coinages were coins of various denominations from different countries. Jazz set those coins aside as the banks will not accept them for deposit; this started her coin collection. Since then, she dreamt of visiting all the places where the coins came from.

Her parents moved her three times in three schools in the first few weeks of her first grade until she settled in and completed her elementary and high school education at the St. Paul College of Manila, where she graduated with honors in 1968. She then completed a degree in Political Science from the University of the Philippines in 1972 then pursued law school. When Martial Law was declared in 1972, classes were suspended for more than a month. Jazz briefly contemplated on shifting to Foreign Service and even took the qualifying exams. However, when her law school classes resumed, she decided to stay on course. Incidentally, she was among the ten percent who were women in their entire law class who graduated from the U.P. College of Law in 1976. Academically astute, she became a member of the prestigious "Order of the Purple Feather," the official honor society of the U.P. Law School. She practiced law for a decade, and because of her love for travel, she always volunteered to handle out-of-town cases, which allowed her to travel for free around the country. Jazz would always schedule her court hearings closer to the weekends so that her weekend can be used to travel around areas where she was working.

At the height of the "People Power Revolution" in 1986, she and her ex-husband decided to move their family to Los Angeles in the United States, as she felt the country was a chaotic place to raise her kids. During those challenging years, they could not afford to travel, as they were starting a new life in another country, with new careers, while raising three young children.

She went on to establish her own business of operating residential care facilities for the elderly in California, called Manila Manor, which she ran for 34 years. As a business owner, it gave her the flexibility to raise her kids while providing for her family. For the next two decades, Jazz devoted her life entirely to her children, her business, and the pursuit of the American Dream for their family.

Her efforts paid off: not only was she able to raise three very successful children who grew up completing their studies in prestigious American schools, namely Georgetown University, Yale University, and Boston College; Jazz was also able to pursue her dreams in her later years.

Her first major travel as a U.S. citizen was a trip to China in 2000 with her mother after her father just passed away. She followed it up in 2002 with her two siblings, their families, and her 83-year old mom when they toured six countries in Europe, namely, Austria, France, Italy, Liechtenstein, Switzerland, and the United Kingdom. Her third trip was in 2003, again with her mother, to Greece, Croatia, and Turkey. Her fourth major travel was in 2004 when she joined the MBA class of Pepperdine University in their China and Tibet trip as a friend of an MBA student. It was a uniquely intense trip with thirty-year-olds and homestays with the locals in rural China where no one spoke English, and she did not speak Chinese.

From 2005 onwards, the year her third child graduated from college, she started making at least two major trips a year. Her destination was determined by what was on sale and when she could travel. Apart from her regular travels, she joined her friends to the annual medical missions hosted by the University of the Philippines Medical Alumni Society in America, which brought them to far-flung rural areas in the Philippines. Thereafter, she traveled four times a year whenever her schedule allowed. She has since reached 128 countries and territories, 90 of which are United Nations-member countries. Later, she was inducted as a full-fledged member of the Travelers' Century Club in Los Angeles in 2017. Jazz loves

> *"They thought I was mad for living my dreams. Likewise, I thought they were mad for dying with theirs." – Daniel Saint*

to take cruises, having sailed around the world, crossing all the seven seas. She has seen all the Seven Wonders of the World. She saves coins and bills from every country she has visited. She also loves tea sets and that has been the item she almost always brings home from every trip.

All told, Jazz always strives for excellence in all fields of endeavor. She hopes that in the twilight years of her life, she would be able to see the whole world before she kicks the bucket.

Jazz's Travel Milestones:

✓ Traveled 128 out of 329 countries and territories (90 out of 193 U.N. countries)

✓ Reached 100th country and territory in 2017

✓ Visited 6th continent in 2008

✓ Completed trips to all new 7 Wonders of the World in 2015

Ramchand "Rambi" Francisco

Great Pyramid of Giza, Cairo, Egypt

- 📍 **Philippine Hometown:** Las Pinas, Metro Manila
- 🏠 **Primary Residence:** Los Angeles, California, USA
- 💼 **Profession:** US Air Force Service Member
- 😎 **Age:** 35
- ✈ **Traveled Continents:** 6
- 🗺 **Traveled Countries & Territories:** 104 (72 U.N.)
- 🌐 **Travel Site:** https://travelhackrr.com

Rambi was born and raised in the Philippines, where he drew his love for travel at an early age from his parents, who were avid travelers themselves. After spending his formative years in San Beda Alabang (formerly St. Benedict College), he studied Electronics Communications Engineering at the De La Salle University with a full scholarship. His first plane ride was when he immigrated to the United States when he was 19. After a few months in Los Angeles, he decided to join the United States Air Force (USAF). During his time in the service, he continued his education and received his MBA degree from the University of Maryland Global College. He has also completed his coursework for his Doctorate Degree in Business Administration-Leadership at Walden University in 2021.

He has been serving in the USAF for the last fourteen years, where he has been stationed, deployed, and has performed military and humanitarian operations in multiple countries around the globe. Throughout his military career, Rambi became a bemedalled soldier, earning the USAF Commendation and Achievement Medals, Global War on Terrorism Expeditionary and Service Medals, and the distinguished John L. Levitow Leadership Award. Moreover, he has garnered the Airman and Non-Commissioned Officer of the Year Awards no less than four times. Rambi is a proud Master Mason and a member of the American-Canadian Grand Lodge. He is also one of the co-founders of the Philippine Global Explorers travel community, where he is a Founding Director and the Charter Vice President of Internal Affairs.

For most of his travels, he is accompanied by his wife, Racsianne, who has been to eighty-four countries and territories, as well as their three young daughters. Their oldest girl, who turned five years old in 2020, has already been to fifty-five countries and territories! With his family in tow, one of their crazier trips was when they completed twenty-seven flights that covered twenty countries within thirty days. Rambi has lived and worked in many countries, each for at least half a year, including the Philippines, USA, Iraq, Kyrgyzstan, South Korea, Italy, Israel, and Germany.

Rambi's love of travel went into overdrive in 2011 after reading an article about a guy who completed all one-hundred and ninety-three United Nations member-countries, merely with the use of travel miles and points. In only nine years, Rambi managed to reach over a hundred countries and territories. An aviation geek obsessed with the Airbus 380, he has flown premium cabins on A380's operated by Asiana, China Southern, Etihad Airways, Emirates, Korean Air, Malaysian Airlines, Singapore Airlines, Thai Airways, and Qatar Airways. Rambi is also a big sports nut, having been to the Winter Olympics, FIFA Football World Cup Finals, and NBA Finals.

Rambi utilizes "travel hacking" – which involves working within the existing rules set up by airlines, credit cards, and hotels, and using them to one's advantage to earn free travel – to fund a big part of his journey, which allowed him to treat his parents, other family members, and friends to trips to Europe, the United States, the Middle East, and Asia simply with the use of miles and points. Back in 2016, he attended the Frequent Traveler University (FTU) in Las Vegas with other miles-and-points enthusiasts from around the world. In 2021, Rambi also attended multiple FTU virtual conferences, meeting miles-and-points hobbyists from around the globe. Rambi is also an active member of the Travelers' Century Club (TCC) and has attended conferences in Brussels, Belgium, as well as in Stuttgart and Nuremberg, Germany. With the help of "travel hacking," he has earned and burned over fifteen million miles and points. He has managed to take over a hundred flights in first and business class for free, as well as over one hundred hotel nights for free with brands, such as Conrad, Waldorf Astoria, Park Hyatt, and Aman Resorts. Because of his expertise, Rambi has been invited to conduct Travel Hacking Masterclass workshops by multiple travel groups. He freely shares his knowledge and has successfully helped over a hundred students with their travel goals.

Among his favorite activities, two of his frontrunners are scuba diving and food tripping He has logged dives all over the world, including in the Maldives, Bora-Bora, Seychelles, Israel, Indonesia, and Thailand, to name a few. He and his wife also love going on food tours and seek out everything food-related, from the most well-known street food hawker stall to the best Michelin-star dine-in restaurants they can find.

A certified thrill-seeker, Rambi's favorite activities comprise of: skydiving in Germany; World War II wreck diving in Coron; driving in downtown Baghdad; Great White Shark Cage Diving in South Africa; para-gliding in South Korea, Canyon bungee-jumping in Bohol, para-sailing in Thailand; white-water rafting in Alaska; as well as Manta Ray and Eagle Ray expeditions in Bora-Bora, just to name a few.

> *"A journey, after all, neither begins in the instant we set out, nor ends when we have reached our door step once again. It starts much earlier and is really never over, because the film of memory continues running on inside of us long after we have come to a physical standstill. Indeed, there exists something like a contagion of travel, and the disease is essentially incurable." – Ryszard Kapuściński*

In 2017, Rambi competed in a military version of the Amazing Race in Seoul, South Korea, where his team won the competition against nine other teams.

Rambi has a driven personality. Whenever he commits to something, he goes all in and gives it his all, believing in maximizing his full potential and the never-ending pursuit of his best self. After his service in the military, he intends to retire and move back to the Philippines before the age of forty, where his dream is to own a small resort by the beach. He loves sharing his knowledge, experience, and blessings with others; and believes travel is a great tool where he can do all that! Rambi lives by the mantra, *"Live your life to the fullest, for tomorrow is not promised!"*

Rambi's Travel Milestones:

✓ Traveled 104 out of 329 countries and territories (72 out of 193 U.N. countries)

✓ Reached 100th country & territory in 2018

✓ Visited 6th continent in 2017

✓ Earned and burned over 15 million miles and points on flights and hotels

PART 2:

TRAVEL STORIES

CHAPTER 3

EXCITING AND HAPPY STORIES

What are the most adventurous things that you have done while traveling?

"Embarking on a nine-month expedition through Africa is a journey filled with superlatives from start to finish. Having traveled through 34 countries and territories, covering a distance of approximately 42,000 kilometers, it was the longest, toughest and most extensive journey of this kind I've ever done in my life. It was in this continent where I had lived through the most inhospitable places and witnessed the most extreme poverty; yet it is also where I had been mesmerized by the most stunning landscapes and the most beautiful people, and had experienced the most thrilling adventures. Indeed, this continent delivers unique experiences to travelers like no other.

This journey kept me physically active and constantly on the move. Riding on the dunes of the Namib Desert, river-rafting the treacherous waters of Uganda, wading in the vast cool rivers and lakes of Guinea and Ghana, hiking up the Todra Gorge in Morocco, and embarking on a sight-running trail through many villages and cities throughout the continent were some of the adventures that had challenged me physically and emotionally. As a middle-aged woman, I was as physically capable in rising above these challenges, as I had been during my college years, when I was a varsity player. That said, there were many aspects of this expedition which have been far more grueling. Plodding in the most desolate deserts of the Sudan in the midst of scorching heat, trekking the deep, steamy jungles of Cameroon, and enduring hours-upon- hours of dirt-road road trips inescapably took their toll.

My solo treks have emboldened me to be more adventurous. Occasionally, I would abruptly jump off a vehicle and leave my travel group to visit places not included in our itinerary, just to embark on a journey of self-discovery. Being a natural introvert, I often find great solace by being by myself. And believe it or not, interesting encounters resulted from some of these solitary jump-offs. Case in point: in my trips to Johannesburg, Eswatini (formerly, Swaziland), and Mozambique, I learned to use public transportation; explored places unattended; did not hesitate to ask for help when I needed it; trusted my instincts; and became increasingly confident and independent.

I foolishly chased adventure. I sought adventure, even knowing that I could be killed in the pursuits. In Cameroon, I found myself being led by pygmy guides and armed forest rangers in the deep of the jungle, then confronted and growled at by its wild inhabitants – the lowland gorillas. The mountain gorillas I was going to see in Rwanda were tame and habituated; they weren't going to be good enough. Crossing the border into Gabon, I signed up to be a passenger of a beaten up and overloaded pickup truck; then its worn tire exploded predictably in the midst of our journey. It's all part of the African experience, I told myself. Admittedly, I made selfish and unwise decisions knowing I had loved ones, my Dad and (then) husband in particular, who were worried about me taking this journey and would be devastated if I got killed. At that time, I felt this is what I signed up for. I wanted the adventure and the space to experiment with life. I wanted to be shaken up." — **Riza**

*　*　*

"I decided pretty much on a whim to change plans and visit Gibraltar — from Malaga on my way to Cadiz — and then hike up the Rock, a monolithic limestone promontory. For its scenic views alone, the last minute side trip was worth it. Crossing the highway back from Gibraltar to Spain side I was flagged down by the Spanish police for lingering on the highway — yes, the yes, the very highway (with a pedestrian lane) that doubles as an airport runway in Gibraltar. Apparently, the gates were already down, with an oncoming plane about to land. Because I was so preoccupied with taking selfies, I did not hear the police frantically whistling at me. I ran as fast as I could, and later, got 'reprimanded' by the friendly police - who also told me I wasn't the first tourist to have committed such infraction!" — **Raoul**

* * *

"I traveled by land from Bogota, Colombia all the way to Sucre, Bolivia (approximately 6,000km), covering four huge countries in a span of four months. I only decided to take a flight from Sucre to Santa Cruz because there was a huge landslide blocking the highway which would have hindered my passage. I resumed my trek on public buses from Santa Cruz all the way to Rio de Janeiro with a huge detour to Foz de Iquazu. Highlights of the trip included running a half-marathon in Quito (3,000 meters above sea level), summiting Huayna Potosi (6,088 meters above sea level), Santa Cruz trek in Peru, Machu Picchu, Pailon del Diablo in Banos, Rio de Janeiro, Salar de Uyuni, and Nazca Lines." — **Jon**

*   *   *

"My photo had a tagline, 'Far away from it all.' How difficult was it to get to that island? Robinson Crusoe Island is also known as Juan Fernandez Islands. The trip was a private jet experience from south of Santiago International Airport. I planned to embark on this adventure twice. It took more than six months of waiting for flights with Aerocardal. Their flights operate only from mid-November to early March. After this, the season was closed. Flights were only once a week. And depending on the wind factor, the time of stay would either be cut short or be extended to several days. Horizontal winds or cross-winds were prohibitive for aircrafts to take off or land.

I arrived in Santiago two days before the scheduled flight. As soon as I arrived, I had to call Aerocardal Airlines, check-in and give my hotel info. My airline ticket had a mini-map of the place that I would check-in. "Easy," I thought. From the sketch, it looked like I could walk from the international airport to the domestic airport. But... *"Uh-oh!"* It turned out that the latter was located very, very far back; the Uber driver even had to call Aerocardal and ask how to get there. After a 30-minute ride, we finally found it. It turned out that the "airport" was a small, isolated office that housed private jets!

There were strict regulations on the luggage weight limit. I had to stand with my baggage in tow on the scale three times to ensure that I wasn't overweight! After I signed all the paperwork, I was given a seat number. In fairness, the waiting area was super clean. A generous selection of free Danish pastries, muffins, juice, milk cereal, chocolate candies and fruits were available for the picking! But the weather was not showing a good sign, so we waited a bit more. By 1:00 P.M., when the climate didn't seem to be cooperating, we were all asked to go home. We could try again tomorrow.

They had a limo that took me back to my Airbnb abode. The following day, I was picked up by the same limo and had to go through the same routine of weighing-in and seat assignments, and stick around for the right visibility factor. Unfortunately, I was sent home for the second time. On the third day, I was already getting frustrated but, alas, I did not give up. Armed with perseverance — well, what do you know? Cheers erupted when we were called to board the flight.

The plane we boarded turned out to be a private jet with a seating capacity for 12 persons. Reclining seats, a bag of goodies, snacks, sandwiches and drinks greeted us as we settled on our seats. Excited passengers took pictures, slapped high fives and extended handshakes as we found our way to our chairs. There were six men from Brazil who were on a fishing expedition. The pilot and co-pilot assured us that the weather was going to be cooperative. The flight would be approximately 3 hours, they said, but the plane could only hold so much gasoline for fuel. After an hour or so, we stopped to refuel. Curiously, the gas tank was on top of the plane! An old step ladder was used to climb up; the attendant manually pulled the long heavy gasoline hose like a fire hose. After two hours of what seemed to be an interminable wait, we finally arrived at the loading dock where a small immigration room with ten chairs, a rest room, and a vending machine greeted us. There, we waited patiently for entry requirements to be processed. Our passports and boarding passes were finally stamped. The loading dock was definitely out-of-sight!

The ocean waves were crashing on black lava rocks. In the dark lava beaches, hundreds of thousands of fur seals glided in with the waves. The vocal repertoire of countless fur seals — which sounded more like a cacophony of buzzing chain saws — were like music to my ears. They dove through pounding the surf with easy, graceful nonchalance and land smoothly on the coastal rocks before me. It was a phenomenal site. We went crazy taking photos!

Then, we had to ride a boat. The sky was bluer than the usual azure hue, very clear with a few clouds. The cliffs were often shrouded by an enigmatic mist which probably inspired countless writers who were bewitched with Robinson Crusoe. The boat ride took another one and a half hours.

The island's population was 600; the number of cars was 25; the number of motorcycles was 3. No iota of stress or pollution can be felt among its surroundings. No rental cars, no buses, no taxis. I had the best walking tour ever. Occasionally, a delivery truck driver will stop and offer a ride, but I politely say, 'Thank you, it's a beautiful day I would rather walk and explore every corner of the island.'

I walked everywhere. The island was famous for giant lobsters, each weighing more than 8 kilos. I must have feasted on too much lobster and had a slight indigestion.

I had a busy schedule. I even watched a soccer game 'Cumberland vs. Nocturno' — the amazing thing was that the playing field was right beside the ocean and a view of the magnificent hills of Robinson Crusoe (Juan Fernandez) islands. More sceneries. Colorful houses, mini-markets, and Isla Pacifico Eco-Lodge.

Guillermo, my Isla Pacífico Eco-Lodge host pointed to the church, '*La Capilla de Robinson Crusoe*,' and explained its history. As I was walking along the main road, I was greeted with a friendly 'buenos días' from passers-by. I saw a house festooned with colored balloons on the front porch. I went up the steps and introduced myself as a writer and they immediately invited me to join a baptismal party. This was perhaps one of the advantages of traveling solo: many surprises spring up along the way. The mother of the baby boy eagerly recounted the history of the island and why they chose to live in such a remote cay. She quipped with her broken English, 'here no stress, every day is (at) your own pace'." — **Odette**

* * *

"Most of my adventurous stunts occurred during my twenties in New Zealand: base-jumping off Auckland's Sky Tower; rappelling and spelunking in Waitomo; tandem skydiving in Taupo; bungee-jumping in Queenstown. I've also tandem skydived in Namibia, tandem micro-lighted in Zambia; rode an ATV in Tasmania; tandem hang-glided and rappelled in Rio, zip-lined in Costa Rica and Laos, and para-glided in Nepal. I've explored caves, in not only New Zealand, but also in Aruba, Nicaragua, Cuba, Lebanon, and Socotra, Yemen; walked on glaciers in New Zealand and Argentina; and hiked on active volcanoes, namely, Pacaya in Guatemala, Mount Yasur in Vanuatu, Erta Ale in Ethiopia, and Mount Nyiragongo in the Democratic Republic of Congo. I know some may think this is animal cruelty, but I have ridden a carabao in the Philippines, an ostrich in South Africa, Icelandic ponies in Iceland and Brazil, camels in the UAE, Jordan, and Mongolia, and elephants in Zimbabwe, Thailand, India, and Sri Lanka.

I would like to visit every UN country in the world, and now, people think it's adventurous that I have visited countries with active war zones. In fact, I visited all the countries with active war zones in eighteen months: Afghanistan (June 2018), Iraq (September 2018), Somalia (January 2019), Yemen (January and November 2019), Libya (December 2019), and Syria (January 2020)." — **April**

* * *

"Filming in volatile regions such as Syria, Iraq, Yemen. It was a thrill."
— **Kit**

* * *

"I had an 86-day sabbatical stay in Nepal during the winter holiday just months after the devastating earthquake to fulfill my dream of embarking on a self-guided solo hike on Himalayas' three most-famous routes. I also did a self-guided solo visit to Iran via the Persian Gulf ferry from Kuwait, as well as to Israel, while I'm working and residing in the Middle East. I had to plan strategically the inbound and outbound flights to make sure my Israel visit would not be traceable, not to mention to challenge the strictest airport immigration in the world. I had a crazy flight route covering eight countries in three continents for 36 days, including the solo hikes I made in Mount Kilimanjaro and Mount Fuji. I managed to circumnavigate the world, visiting the 3 continents from UAE to America and back to Asia during the starting month of Covid-19 pandemic. Other adventurous trips I did were 33-hours self-guided solo trips on the Golden Triangle tour of Incredible India, as well as a solo trip to North Korea and self-guided solo trips in Afghanistan, Timor-Leste and Turkmenistan. I also did a lot of biking in Easter Island (Rapa Nui). But before this, I biked to Bagan (Myanmar), Luxor (Egypt) and to Stonehenge (UK). I've also completed a seven months Latin America solo trip from Mexico down to Central and South America visiting 16 countries from November 2020 to June 2021." — **Andie**

* * *

"I've done so much during my six years of nomadic existence, including my Amazon experience where I slept on a hammock for a week while on a boat from the Tres Fonteras (Peru, Ecuador, and Colombia) to reach Belem, Brazil. I also swam with the penguins in Antarctica, hitchhiking in Patagonia de Chile to a week-long trip in Venezuela.

But some of the coolest trips was when I traveled solo around Africa where I've visited 17 countries. To wrap up my trip, I flew to Djibouti where I received my Somaliland visa within 24 hours.

I made some research on how to cross the border, one travel company offers US$500/day which included a private car and a bodyguard. I find it very expensive so I tried to look for a public transport. With the help of my guide, Osman, we negotiated with a Somali guy who is the owner of the pickup truck going to Somaliland. I paid 12,000 Djiboutian Francs (roughly about US$68; I paid a little more so I could get the front seat) with no receipt, just pure trust! Initially, I was hesitant about it, but he felt bad that I didn't trust him so I just let it go. I just crossed my fingers, hoping he wouldn't try to get away with my money. *Ha-ha!* Thank goodness I had Osman with me, otherwise, it was impossible to find this terminal. There were no signages of it at all!

The 13-hour journey started at 4:30 P.M. (that was the only schedule they had). We rode on a pickup car with improvised seats at the back — just a short 20-minute ride to the Somaliland border from Djibouti. I was with Somalis and Djiboutians in the car, one of them was Abde-Aziz, the only guy who spoke English.

When we arrived at the border, I was interrogated for more than an hour at the Djibouti Immigration. They asked me so many questions and they wouldn't let me out. I had to show them all my visas and prove them that I have traveled extensively before and I know what I'm doing. I actually thought that I wouldn't make it to the border because they have declared recently that the border is only for Djiboutians and Somalis (maybe not true!). But lo and behold,

after more than an hour of uncertainty, they finally let me out! The immigration office was already closed when they stamped my passport with the condition that I wouldn't come back to that border because they will not let me in. And he wasn't joking... he was very firm."
— **Kach**

* * *

"In 2001, my travel buddy planned a trip to Nicaragua and since we had two weeks, we could visit other neighboring tropical countries in Central America. We visited as much as we can, we went to see some Mayan ruins, beaches, volcanoes, colonial sites, and enjoyed our stay. On the day we were supposed to go to Nicaragua, when we got to the airport, we realized that the plane was just a six-seater. We were both shocked and went to inquire and they said that it was our plane which would be leaving soon. Sure enough, they advised that we had to be weighed in order to get our seats. We got on and took off; but I was really frightened. I had never been on such a tiny plane; the two-hour flight left me speculating 'what if' something happened to the flight, we would be lost in the jungle. Fortunately, the little plane landed safely into Nicaragua. That was more like an adventure."
— **Luisa**

* * *

"Ooh, where to start? There's the skydiving experience in one of the biggest drop zones in Europe. World War II wreck diving in Coron, Palawan and deep-sea diving in the Maldives. There was also canyon bungee-jumping in Bohol, Philippines.

Manta and eagle ray scuba expeditions in Bora-Bora. Jumping off a mountain and paragliding in South Korea. Parasailing in Phuket, Thailand. But perhaps the most memorable would have to be the cage-diving experience with great white sharks in Gansbaai, South Africa. As an avid scuba diver, this was something I will never forget. I was fortunate to have witnessed three of those behemoths up close."
— **Rambi**

* * *

"Back in 2013, my husband and I traveled two months in East Africa, visiting Kenya, Uganda and Rwanda. One of our highlights was the trip to Bwindi National Park but the access roads were flooded due to low pressure. Going there was hard because of recurring landslides, so we pushed the car together with the locals, and had to walk in the mud barefoot. At the end of the day, we found a tent in the middle of nowhere, it was still raining cats-and- dogs. The next day, after visiting the national park and it was time to go back to the city, the situation got much worse: while attempting to cross a landslide, the car got damaged and all of us were hopeless but, as if on cue, a group of local boys appeared out of nowhere from the bushes and helped us fix the car. When we finally arrived safely in the city, we immediately we went for the car wash. We continued our trek via a public bus to Kigali, Rwanda." — **Vhang**

* * *

"While in Central Asia, my husband and I walked from the borders of Uzbekistan to Tajikistan. We maneuvered our way while countless of trucks transporting goods, lining up as far as you can see and not a single soul in sight except for the two of us. At immigration, the police checked everything in our luggage which took forever. After that harrowing experience, we had to wait for half-a-day because our 'guide/driver' forgot to pick us up!" — **Henna**

* * *

"I was forced into joining a ski group in Vail, Denver, Colorado, although I have not received any skiing lesson before. I had to crawl from the top of the mountain until the very bottom without any outside help, causing me to nearly contract frostbite." — **Jimmy**

* * *

"I've done a lot of adventurous stuff during my global trips, most of which have been daring, such as bungee jumping and skydiving at Queenstown in New Zealand; great white shark cage diving in South Africa; abseiling in Bled; zip-lining in the Laos jungle to see the gibbons; and tubing in Vang Vieng. I've been lucky to have been part of many "adrenaline rush" experiences, such as, trekking the Inca Trail, game drive at Kruger National Park and Ngorongoro Crater, riding a car at 240 km/hour along the autobahn, flying in a wind tunnel, sailing at Whitsundays and Bay of Islands, riding the hot-air balloon in Myanmar's Bagan temples on New Year's Day, flying in a glider in Hawaii, visiting the radioactive Chernobyl nuclear plant and Pripyat ghost city in Ukraine, riding a quad bike and buggy in the sand dunes of Dubai, as well as sandboarding in the Atacama desert.

> *"Each day, there is a chance you might die. And there's nothing wrong with that. Every living being on Earth is facing that same existential rift." – Alex Honnold*

My most dangerous adventures were those that didn't have any safety guarantees, often throwing caution to the wind and relying purely on my common sense for my survival. My mother probably would never have permitted me mountain biking the Death Road in Bolivia, running with the bulls in Pamplona, jumping to the water from cliffs in Hawaii's Big island, swim underwater to enter the Sawa-I-Lau cave in Fiji's Blue Lagoon (and I wasn't even a good swimmer), and trekking to the crater of Mount Yasur, an active volcano in Vanuatu."
— **Dondon**

* * *

"Riding a *lampiitaw* (small *bangka*) from Claveria port Cagayan to Calayan Island. Six hours of non-stop riding the waves until one of the Katig (balancer) broke down on the middle of the ocean with an eight-to-nine-foot wave hampering the *bangkas* (*lampitaw*). The crew were very alert and were valiantly trying to fix the *bangka* while being hit by big waves. We survived with around thirty-to-forty passengers aboard the *bangka*." — **Badong**

* * *

"Sulu was indeed an island paradise. It's so unfortunate that the security situation there won't allow tourist arrivals for the meantime. But my curiosity pushed me to visit the island and see what was there. In 2010, you could get to Jolo, Sulu from Zamboanga City since ferries left the Zamboanga Port every evening and arrived in Jolo at about four in the morning.

I kept my trip to Sulu under wraps, knowing that everyone would dissuade me, except for hardcore travelers, of course! But before daring to set foot on this beautiful yet precarious island, I had to make sure that all bases were covered. In 2010, it was not a good idea to visit the island if you did not have a local to accompany you. As they say, "don't try this at home, folks" since visitors definitely stand out from the language difference alone.

Weeks before, I had been in contact with a fraternity brother. His father, the late Ismael 'Pochong' Abubakar, Jr. was the first Speaker of the ARMM Regional Legislative Assembly. They had graciously assisted me when I visited Tawi-Tawi in 2009. 'Ka Pochong' had asked his cousins in Jolo, Sulu to take care of me while I was there.

It was four in the morning. The M.V. Kristel Jane 3 had just docked in Jolo, Sulu. Boats from Zamboanga usually arrived in Sulu around this time. When I bought my ticket, I was advised to stay in my cabin until the sun came out. In fact, the lady at the counter said to me after handing over my ticket, 'Good luck!' Such were her words of encouragement which, of course, had an ominous twist of irony and sarcasm in them. Good luck, indeed!

It was still dark when we stepped out of the boat. As we exited the port for our predawn breakfast, melodious chants blared from loudspeakers atop the minarets of mosques around town, piercing the morning silence as muezzins recited the 'Adhan' or Islamic call to prayer.

"So I find words I never thought to speak, In streets I never thought I should revisit, When I left my body on a distant shore." – T.S. Eliot

Jolo was once a charming town. It used to be a walled city during the Spanish colonial period. But there was nothing much left to remind us of its fortifications, save for a few bricks and watch towers hidden by the urban chaos that politicians left unregulated. My host lamented the destruction of the historic wall that formed an inherent part of Jolo's heritage. Add to that, Jolo was totally destroyed in 1974 as a result of heavy fighting between government forces and the Moro National Liberation Front (MNLF).

Breakfast was *satti*, a dish composed of small pieces of beef grilled on skewers and served submerged in a bowl of sweet and spicy sauce. Included in the bowl were pieces of *puso*, rice that is cooked inside a palm leaf pouch. In Malay, it's called 'ketupat.' They also served grilled chicken.

Satti was actually a dish native to Jolo, Sulu. The ones in Zamboanga, in fact, originated from Jolo. The hawker stalls were abuzz with activity so early in the morning since it was the fasting month of Ramadan.

After breakfast, we proceeded to the house of my host where I took a quick nap. After a good rest, we proceeded to explore Jolo and the neighboring towns of Patikul and Indanan.

We first dropped by the Sulu Provincial Capitol in Jolo. I noticed that the centuries-old trees that lined the avenue led to the Sulu Capitol. How I wished many of our old cities were able to preserve their trees. We could also see the vernacular architecture hidden under the urban chaos of Jolo. If only the politicians there had the political will to clean up the city and preserve its character, Jolo would have been an even more fascinating town.

Later in the afternoon, we motored to the town of Indanan to visit Bud Datu where the grave of Raja Baginda, the first Muslim ruler of Sulu, was located. We had to enter a military camp to reach it. And since it's in a military camp, it's relatively well-maintained. We could also see a nice panoramic view of Jolo from Bud Datu. But unfortunately, as we were walking from the Rajah Baginda Shrine to our vehicle, the sun disappeared again and it started to rain. So we ditched the view.

We had to wait for the rain to go again before proceeding to our last stop, said to be one of the best beaches in the Philippines with a wide expanse of white sand that could rival that of Boracay. It was about twelve kilometers from Jolo in the town of Patikul. Quezon Beach was highly-recommended with a caveat though: we may need a security escort to visit.

My hosts didn't mention any of that so we proceeded to Quezon Beach. We passed by several military camps and checkpoints along the road that went deeper into Patikul. I marveled at the houses, which were very fine examples of vernacular architecture. They rarely used hollow blocks in Patikul. And I felt I entered a time warp as we drove through since these could have been the architecture of Maynilad when the Spaniards arrived there almost five centuries ago. Most of the wooden houses were elevated on stilts with covered porches on two sides. An elevated walkway connected the main house to another structure behind the house which served as a kitchen and cleaning area.

We finally reached Barangay Igasan and we parked along the beach. There wasn't too much sun. But I could see that the beach was stunning even with the overcast skies. The beach was wide and the water was baby blue. I was told that further down the road, the sand was even finer, powder fine, in fact, that could rival the best beaches in the country. But I had to save it for another day. I did not want to push my luck any further since it was starting to get dark. The sun and sand would have been an impeccable combination. But I guess I'd have to wait for another trip, hopefully when the situation became a bit better.

What I liked about the beach was that the houses were all made of native materials. So, it really gave me that tropical feel. At least for now, it will stay that way. But I wonder how they would manage development there if the situation got better.

Back in Jolo, we had dinner and I got to try more of their local meat dishes, such as *pastir* and *pyesak*. Relatives of my host, curious as to where I went, asked which places

I visited. When I told them that we came from Quezon Beach, I got startled reactions. One even asked my host if they really brought me there and said I was brave to even visit. It was only then that I found out that the area was where many of the kidnappings that year occurred.

After dinner, I was brought to the Jolo Port to catch the 8 P.M. ferry back to Zamboanga. It was the M.V. Kristel Jane 3 which ferried us back to Zamboanga; curiously, I got myself the same cabin room. I was back in Zamboanga City at 4 A.M. just in time for another *satti* breakfast. As soon as I was done, I rushed to the transport terminal in Guiwan to catch a bus to Pagadian." — **Ivan**

What was the craziest thing you've ever done in your travels?

"One of the most daring things I've ever done on my world travels was having to stand in the Kjeragbolten boulder located in Kjerag mountain in Rogaland, Norway. Getting there was quite an effort with the three-hour drive from Stavanger and then the two-hour hike along with various flora and fauna, which vividly reminded me of 'The Lord of the Rings.' It was unbelievably daunting at first to see this massive boulder lodged between the mountains, suspended above an almost thousand meters deep abyss. I knew I couldn't let the opportunity to be on top of it pass, so I made my way slowly and carefully on top of the rock. It was a very cold and windy day that day, and I could feel the strong winds blowing against me as I was standing there. I had to stand still because with one wrong move, and I'm gone. The toughest part was getting back with the tiny surface area to grip on and keeping myself balanced. It's definitely not for those with a fear of heights. Those who dare win!" — **Dondon**

* * *

"Walking with the lions in Fathala wildlife Reserve in Gambia was a once-in-a-lifetime experience!" — **Luisa**

* * *

"I visited Luján Zoo in Argentina and went inside a lion cage. Prior to entering the lion cage, I had to sign a waiver stating that I'm going at my own risk and if something bad happens, not even the trainer will be responsible. Despite this, I still pushed through with it since one of my bucket lists was to touch a lion." — **Badong**

* * *

"There were plenty, such as (1) traversing the African continent for nine months in a re-purposed military truck, sleeping in tents and jumping off the truck on occasions to explore other parts of the continent by myself; (2) visiting war-torn countries and countries still at war like Syria, Somalia, Libya, Afghanistan and South Sudan; (3) visiting Mecca and Medina in Saudi Arabia, both off-limits to non-Muslims but the implementation of liberal reforms by the current government had enabled me to make a special visit." — **Riza**

* * *

"When I visited Svalvard, Norway, I joined the racing competition of snow mobiles where my bravado caused me to fall to a deep crevice. Luckily, I was retrieved with numerous sprained bones and I was able to go back home to the Philippines. Incidentally, this accident happened during a Good Friday." — **Jimmy**

* * *

"Not really crazy, but more spontaneous — while in Morocco in the early 2000s, I had no set itinerary. In Marrakesh, I woke up one day and decided to visit a seaside city called Essaouira by bus. At the bus station, I boarded a bus with five other people in their early twenties who were going to Essaouira to help organize the biggest musical festival of the year — the Gnaoua World Music Festival. One member of the group caught my eye, so I chatted her up during the bus ride. When they found out I was just planning on visiting for a day or two, they asked me if I wanted to stay with them in a house that they had rented for the next 10 days, and if I wanted to help organize the festival. I took a quick glance at my beautiful new friend and, of course, said yes! As someone who had no previous knowledge of African music, it was an eye opening, fun, and exhausting experience, and we all transformed from random strangers to great friends by the end of it all!" — **Brian**

*   *   *

"Quarrelling with a person who was questioning me and my boyfriend on why we didn't sit next to each other on the bus. We didn't sit together because we wanted to have more personal space, assuming of course, that no one occupied the seat next to us if people traveled by twos. We always took the risk of assuming that. So, if there was a person who traveled by himself, we also risked the chance of sitting with a stranger." — **Jazz**

*   *   *

"I had a few itineraries that may have sounded crazy at first but which I'd thoroughly enjoyed while doing it. One was a ten-day trip that brought me from Amsterdam to Warsaw to Krakow (3 nights) to Athens (6 hours for a quick lunch at Grand Bretagne) to Istanbul to Manila (10 hours) to Singapore (9 hours) to Kuala Lumpur Airport (quick transit) to Jakarta (3 nights) back to Manila (16 hours) to Istanbul, and finally back to Amsterdam.

When I was working in Bandung, Indonesia, I had to get out of the country one weekend every month because of my stay limitations. I had three flight back to Philippines where I would only spend twelve hours in Manila — just about enough time to return to my place, unpacked to drop off souvenirs or food stuffs bought from Indonesia, have lunch or coffee with friends, got a few hours of sleep before returning to the Ninoy Aquino International Airport to catch my flight back to Indonesia. I did this twice when I was working in Australia and had to saunter briefly to Singapore so I wouldn't overstay, spent eight hours buying supplies or clothes because shopping in Singapore was so much cheaper, got a meal or two to satisfy my Singaporean food cravings, and ran back to the airport to catch another flight back to Melbourne." — **Jon**

* * *

"My craziest would have to be the round-the-world trip in 2018 where my wife and I visited twenty countries and took twenty-seven flights in thirty days. All these with our twenty-month old daughter in tow. Whoever said that traveling stops or slows down when you have kids hasn't met us yet." — **Rambi**

* * *

"I think some people would say that the craziest thing I've done was when I quit my full-time job to travel around the world and live a nomadic lifestyle from 2013 until 2019. But maybe the craziest was in April 2017, when my (then) husband and I decided to buy a 37-foot sailboat in Florida and we both didn't have the experience of sailing a boat on our own. We refitted this 1971 monohull boat and enjoyed our two years of having a liveaboard life where we sailed in Florida, the Bahamas, the Turks and Caicos Islands, Dominican Republic, and Puerto Rico with our two cats." — **Kach**

* * *

"Negril Beach, Jamaica. I was very young then. My friend Annie and I commissioned a glass bottomed boat and asked the guide to bring us to a secluded location of the nude beach. Some called it hedonism, others found it offensive, but in Europe it's almost everywhere, even in Barcelona. It's all about what others think. People did not even stare, we all laid on our mats under the shade of swaying palm trees. I also went nude swimming at Saint Maarten island (French side) of the Caribbean islands." — **Odette**

> *"There is no moment of delight in any pilgrimage like the beginning of it." – Charles Dudley Warner*

Have you done some trekking and hiking? If so, which ones have been the most challenging?

"When my husband and I did the Annapurna five-day trek in Pokhara, Nepal, we met some Maoist rebels. I was with three other trekkers who were all Americans. The Maoists, showing off their guns, demanded money as 'permit' (FYI — we already got government permits prior to the hike). I smirked at them and sort of said something like, "Aren't you shaming your country to the world? Huge portion of your government's source of income is tourism." It later dawned on me that these militants had guns after I had said my piece, but they were young (one was a lady who looked no older than eighteen). I said, "I do not have money," then quickly shut my mouth. But they still demanded US$10 from the Americans and US$5 for me. We were all given fake permits. And we had ours preserved to this day." — **Henna**

* * *

"Mount Kinabalu is the rooftop of Borneo and the most prominent peak in Southeast Asia. Words cannot describe how I felt as I stood on Low's Peak, the highest point of Mount Kinabalu at 4,095 meters above sea-level. It was a challenge to get up there, an ordeal even. But the view from the top was nothing short of breathtaking.

After a previously unsuccessful attempt to reach the summit of Mount Kinabalu in 2010, I finally conquered it in 2011. That year was extraordinarily difficult. Although the weather was expected to be good, the La Niña made everything so unpredictable.

135

On the way up to the Laban Rata Rest House, I had to deal with torrential rains, as well as the slippery and muddy trail it created. Just like the previous year, I slowly inched my way up the steep six-kilometer route to Laban Rata for eight grueling hours; this was considered a personal feat, given the fact that I was not as physically fit as I should be.

There were two trails that led up to the summit: the Timpohon and the Mesilau Trails. The Timpohon Trail was the shorter and more popular route. It was approximately six kilometers away from Laban Rata, and an additional 2.72 kilometers to the summit which was known as Low's Peak, for a total of 8.72 kilometers. The Mesilau Trail was 1.6 kilometers longer and navigating it was more of a challenge, since the pathway was not as developed as Timpohon. The two trails converged after Kilometer 4 in Timpohon.

At the Timpohon Gate (1,866.4 meters above sea level), the guide conducted one final briefing before we were allowed to enter the gate. He then submitted a record to the ranger situated at the gate, who then asked us to sign beside our names and check that we were carrying the correct permit. From there, the grueling six-kilometer trek up Laban Rata began, with the nearest accommodation facilities nestled around 3,272 meters ahead in Mount Kinabalu.

That evening, amid the heavy downpour, our 30-member expeditionary party prayed for a miracle — that the skies would clear up the next day — in time for our assault to the summit. Our prayers were answered. We were gifted with a very beautiful morning with clear blue skies.

It was another 2.7-kilometer trek to the summit. Our group left Laban Rata a few minutes before 3 A.M. and nearly missed the cut-off point at the Sayat-Sayat Hut. We needed to reach the checkpoint at 5 A.M. But thank God we were allowed to continue, despite having arrived a few minutes late since the weather was relatively good.

Despite feeling weak and having to endure the biting cold winds, the lure of the summit's grandeur, as well as a lot of nudging from my friends, helped me inch my way to the top. I got to enjoy the view from Low's Peak for quite a while. In fact, I had the summit all to myself since me and my guide were the last ones to make our way down.

As if reminding us that the clear morning was simply a prayer answered, it started to drizzle again as we descended from the summit. Then, the drizzle turned into a light rain. The rocks started to get really slippery. And there were portions of the trail where I had to hang on for dear life as I maneuvered through a steep cliff. And then, just when we thought things were already bad, the sky opened its floodgates and released a torrential downpour that transformed the trail into a cascading stream of mud. So, we had to deal with that from the Laban Rata Rest House all the way down to the Timpohon Gate. No doubt, those few precious minutes when the heavens opened were well worth the effort." — **Ivan**

*   *   *

"I always looked forward to mountain trekking expeditions simply because I love to experience the great outdoors, especially with a stunning view. Despite being an urbanite, I easily get bored in museums, malls, and other such city sightseeing attractions. But just like any average enthusiast who call themselves as 'fun-hikers,' I don't physically prepare for major hikes. Oftentimes, I easily experience bouts of painful leg cramps on my knees and calves just by walking a little more than five kilometers, even on a paved road. Being a late-bloomer in the trekking scene, I embarked on my first hike was when I was already 27 years old, when my colleague, who was also a novice in a trekking group, asked me to accompany him. After that rather pleasant experience, I joined with the same group in more hiking expeditions, before I finally decided to organize my own hiking group. Our first group expedition began at the third-highest mountain in the Philippines during the country's wettest month, with majority of the participants being strangers who joined through an online travel forum.

As a casual hiker, I classify the type of mountain trekking challenges based on the level of difficulty. My most locally challenging climb had to be the day-hike extended traverse of Mount Mariveles (a three-hour road trip from Manila; the mountain trail extended from Pantingan Peak to Tarak Ridge) that took me all of twenty-two punishing hours to complete. I considered the

Mount Mariveles Range on 'die-hike' as my most knee-wrecking adventure due to: (1) the mountain's savage mossy forest with steep, slippery and complex trails on soil, rocks, laid trunks and dead trees; (2) lots of *"limatik"* (blood leeches) from Pantingan Peak assault to El Saco Peak descent. I had three bouts of *"limatik"* attacks that struck my lower leg and face; (3) wet-and-wild adventure; it rained three times since 11 A.M., even my underwear and socks were already soaking wet only around 14-hour hiking with a cake-full of mud that crept up my shoes; (4) insufficient water source and packed-food rations, due to my very limited knowledge about the trail. In fact, my usual provision of 1 liter of water and few breads for a day-hike was undoubtedly in very short supply; and, (5) I already exceeded the endurance limit of both of my knees; I cannot even lift my left leg properly during my descent and cramps on both legs during my ascent, while being too tired and sleepy with very short bouts of rest.

I didn't have much issues in my expeditions abroad, as I already embarked on a lot of successful sets of major hikes, most of them self-guided solo climbs. For my longest day on a trail adventure, however, I considered the self-guided solo hike on the Himalayas' three famous routes (Everest Basecamp / Kalapathar Peak via Jiri, Annapurna Circuit and Lantang Valley) as my most challenging. I conducted these hikes around eight months after Nepal was devastated by a magnitude 8.0 earthquake. I could just imagine what the locals must have gone through psychologically and financially. Among those three trails, the most rewarding was Everest Basecamp / Kalapathar Peak via Jiri due to the trail's daily sinusoidal ups-and-downs of elevation gain. I chose the unusually circuitous route in the Everest Base camp during off-peak season because somehow, I wanted to help by means of visiting the affected villages of the earthquake. Moreover, I wanted to chart my own version of the Himalayas trekking adventure, as seen in the movies 'Secret Life of Walter Mitty' and 'Everest.' Yes, I did reach the target base camps and summits in all of these three trails successfully.

My most challenging rides going to the mountain base were the Jebel Shams in Oman and Mount Ramelau in Timor Leste (both being their respective countries' highest mountains which I hiked in 2017). In both instances, I had to hitchhike on my way to the jump-off point and they were exceedingly far to the main road. Curiously, I decided to give up the trek, not knowing that I was less than a kilometer away to the mountain peak. What made the experience worse was that I lost my trail several times in the middle of the night. What made it scary was the fact that I was all alone with very limited resources.

I regarded the Triple Day Hike (again, a self-guided solo adventure) in Cape Town's Table Mountains Three Peaks (Devil's Peak, Table Mountain and Lion's Head), as one of my most memorable in terms of the number of mountains I was able to hike in one day. I was motivated to do so in order to maximize my time on the area, save on Uber ride fees and challenge my middle-aged self-doubts. Table Mountain was the most difficult trail and the highest among the three peaks. I finished the hike from ascent to descent in approximately 3.5 hours with more than 600-meter altitude gain. There were no registration fees and guides weren't required, since the trails were well-recognizable). But it wasn't a walk-in-the-park either: the trails were steep, rocky, sandy and very much exposed to scorching heat. In embarking on this type of trail, hydration and sun protection were a must. I had only less-than-a-liter of drinking water in my canteen, and I only wore a loose shirt and casual shorts. However, I still overcame the challenge!

Finally, I accomplished the following self-guided solo budget major hikes in Latin America in 2021: (1) Pico del Aguila (3,846 masl) and Cerro del Ajusco (3,951 masl), Mexico City's highest mountains; (2) Paricutin Volcano (2,774 masl) in Mexico, the world's youngest volcano; (3) Tajumulco Volcano (4,220 masl), Central America and Guatemala's highest peak; (4) Acatenango Volcano (3,976 masl), Central America's 3rd highest peak; (5) Colca Canyon in Peru, world's 2nd deepest canyon at 3,400m; (6) Machu Picchu, Peru (my 7th of the Seven New Wonders of the World); (7) Chimborazo Volcano Refuge glacier trail (5,200 masl), Ecuador's highest mountain; (8) Tungurahua Volcano Refuge Camp (3,830 masl), Ecuador's 10th highest mountain and one of South America's hardest dayhike with highest altitude gain at 2,010m; and (9) Quilotoa Volcano Crater Lake Loop 3-Days Hike, Ecuador's best DIY multi-day hike in Andes Mountains." — **Andie**

* * *

"I have done quite a lot of trekking and hiking during my travels. Hiking Everest Base Camp (5,400 m) from Lukla for several days was the most challenging so far." — **Riza**

* * *

"Trekking in Uganda and Central Africa Republic were challenging because it was mountainous, passing through the muddy forest, crossing streams with our feet submerged in water, long hikes before we came in contact with gorillas." — **Luisa**

*   *   *

"I don't usually go trekking or hiking as it's not my style of travel but I've done the Gorilla trekking in Uganda. The most memorable in my book, however, was when I went on a hike in Kheerganga in the Himachal Pradesh region of India to check out the famous "Rainbow Gathering". These are temporary gatherings to which people from the rainbow tribe travel from various places to spend a couple of weeks close to nature. These gatherings take place in the woods far off from the civilizations and share the same ideologies of peace, harmony, love, light, freedom and respect. The lifestyle is hippie and community living!" — **Kach**

*   *   *

"Hiking has always been my Number One travel activity. While planning for a trip, I always looked for mountains that I could possibly trek. The toughest mountain I've done so far was Mount Sicapoo in the northern part of the Philippines. It was a four-day trek that involved several river crossings, steep ascents, open savannah landscapes, especially on the last day that definitely tested my tolerance to extreme heat. I found hiking in tropical countries more challenging than glacier and alpine countries due to the dense trails and prevalence of mountain creatures/insects (leeches, snakes, mosquitoes, etc.)." — **Jon**

*   *   *

"My most challenging treks include the climb to Adams Peak in Sri Lanka and Mount Pulag in Cordillera, Philippines, where in both instances, I was carried via a portable bed mounted on two poles and carried at each end on the shoulders by porters, called a *palanquin*."
— **Jimmy**

* * *

The most challenging trek I've done was Mount Huangshan, located in Southern Anhui province in Eastern China. Yes, there was a cable car first, but then it was intense hiking to Lotus Peak (1864m); it involved climbing up and down thousands of granite steps, very exhausting but incredible scenery. It is considered one of the steepest hikes on earth. The whole park contains more than 77 imposing peaks and is a World Heritage Site since 1990 for its outstanding natural beauty. But we were not alone on that day. It was a people's migration of thousands of local Chinese trekking up the Lotus Peak together with us. Luckily, on the top, there was a small canteen selling basic Chinese food and green tea." — **Vhang**

* * *

"I used to run four miles every day, skip rope, climb our office building of fifty-two floors to stay fit with Sierra Park hikers. Lately, I am not that much of a hiker. But the most memorable places which I hiked for one-to-two day trips were at the Morne Seychelles National Park in Mahe; and the Ordesa National Park as well as the Monte Perdido National Park, the last two are both located in the Pyrenees of Huesca,

Spain. Lush green meadows, enormous forests, and incredible gorges pepper the awesome landscape, made more pleasant with the sound of 'clang, clang, clang' from the bells around the herds of cows crossing the valleys. They were music to my ears." — **Odette**

* * *

"Most of the time in some parts of Alaska hiking alone where I saw one baby bear and one wild deer." — **Badong**

* * *

"It was the 3rd day of our Inca Trail hike in Peru in very rainy conditions. The night before, water got inside my tent, and I will never forget how miserable I was sleeping with soaked socks. Our entire hiking team was so wet and weary, determined to get to the endpoint at the town of Aguas Calientes as soon as possible, that we managed to finish the trek a day earlier than planned!" — **Dondon**

* * *

"Growing up in a big city located in a very flat area of the United States, I wasn't exposed to any hiking activity. A winter break during pharmacy school changed all that, when I decided to embark on my first hiking adventure at the Inca Trail in Peru. Since Chicago is almost at sea level, I quickly found out what altitude sickness felt like within an hour of landing in Cusco. For me, the most challenging aspect of a great hike was not the physicality required, but the increasing altitude. Within that year, I thought I was experienced enough to carry my own tent, food, and backpack on a four-day trek through Tongariro Northern Circuit in New Zealand. Looking back, I realized how inexperienced and naive I was! I carried and placed too much unimportant stuff incorrectly and my tiny body suffered. But I made it! I hiked 50 kilometers by myself, and what a sense of accomplishment that was! When I hiked the four-day Milford Track a week later, I did not find it as challenging as Tongariro because I was already stronger. It was probably just as difficult and just as gorgeous, but my mind and body had already changed within a short week.

I've summited Mount Kilimanjaro in Tanzania and trekked the Annapurna Circuit in Nepal for seventeen days. I've hiked to Mount Everest Base Camp. I've climbed Mount Kinabalu, where my guide said I was the strongest woman he ever guided. My most recent summit was Mount Cameroon in March 2019.

I still don't feel that I am an experienced hiker, but I try to do it everywhere I go. If it is a long hike, I always hire a porter or a guide. As a city girl, I know I do not have the best survival knowledge and I feel much safer when hiking with someone else. I've done some amazing and challenging single-day hikes through the Grand Canyon, Zion, Bryce Canyon, Arches, and North Cascades National Parks in USA; in Ilha Grande, Brazil; through Patagonia; to Cradle Mountain in Tasmania; the Tatras Mountains in Poland; to the crater lake of Santa Ana in El Salvador and Mount Nyiragongo of Democratic Republic of Congo; the Alps of Interlaken, Switzerland, and so many more places. Since hiking the Inca Trail, I've changed the way I used to travel and now have a list of hikes that I would like to do." — **April**

* * *

"The Inca Trail camping / hiking to visit the Machu Picchu, experiencing some high-altitude sickness (*"sorochi"*.) I was lucky to get relief after drinking coca tea for the most part, to alleviate the dizziness." — **Raoul**

* * *

"Climbing to Takstang Lakhang Monastery in Paro, Bhutan was my most memorable because of the altitude and my age when I did it. Macchu Picchu was also particularly striking, because of the high altitude. Finally, Potala, Tibet because of the altitude, plus I suffered from altitude sickness." — **Jazz**

* * *

"I found the Camino de Santiago quite challenging because I did it rather spontaneously and had no time to train or prepare. And it was one of the very few instances wherein I actually overpacked! With such an arduous trek, every milligram counted!" — **Kit**

* * *

"My abridged Camino de Santiago experience was twelve days of up to 35 kilometers per day hiking, which was extremely challenging in the beginning. Every night, a different body part would ache like never before, including muscles I didn't even know existed! The next morning, however, in true Camino miraculous form, I would wake completely refreshed for a new day's trek.

Mount Pulag in Benguet, was also memorable. The hike was not too difficult, but spending the night in a wet and cold tent with frozen rain flying in horizontally, only to wake up to a cloudy morning preventing my brother and I from summiting, will always be an adventurous but not-so-fond memory.

In terms of a single climb, which I decided on just a few hours before embarking on it, climbing Mount Fuji was probably the most intense ten-to-twelve hours of exercise I have ever endured. I tried taking all my frustration out on the sacred mountain countless times by bashing it with my walking stick!" — **Brian**

Describe your best travel buddy.

"Someone like Anthony Bourdain but a little bit more cheerful." — **Kit**

* * *

"I always traveled solo, but on a few occasions, I've met up and traveled with one or two nomads with the same destinations. I've noticed that what I needed to be on a travel group was also what I saw in a fellow travel buddy — flexibility, a sense of self-awareness, give-and-take attitude, consideration for our safety, and most importantly, the common spirit of adventure. And a sense of humor goes a long way." — **Raoul**

* * *

"He or she must be prompt, clean and knowledgeable about the details of the travel." — **Jimmy**

* * *

"My best travel buddy is someone who has a lot of energy, flexible, adaptable, organized and adventurous." — **Riza**

* * *

"Best travel buddy would have to be adventurous enough to try local experiences with me. Someone who would be ready-to-go at a moments' notice. Someone who would endure ten-segment flight itineraries just to visit destinations for a day. And someone who would be able to eat the local cuisine and stay up all night to try out new food spots. I pretty much described my wife who's my forever travel buddy!" — **Rambi**

* * *

"My wife is always game for any adventure and is as excited to experience the world as I am. I am fortunate that she likes what I like and allows me to do the planning based on what sites I think we'd like to see and what activities I think we would like to pursue." — **Brian**

* * *

"My boyfriend is my best travel buddy. He is adventurous, money is not a problem for him, he is not impatient, he allows me to plan the trip and he agrees to everything I plan." — **Jazz**

* * *

"Best travel buddy is my husband who is himself a very active traveler." — **Vhang**

* * *

I've traveled with my (then) husband in over 80 countries and he's pretty cool but we actually have a different style of traveling. I love traveling with my sister as she's organized and we have the same interest of traveling from backpacking, and road trips to any water sport activities." —**Kach**

* * *

"My sister because she knows exactly how I like it when I travel, from planning to photos." — **Dondon**

* * *

"My close friend who knows history and can speak in a variety of languages…" — **Luisa**

* * *

"I usually travel solo or with one person; a close friend who knows that there are times that I would scream aloud in my dreams; quick to use an iPod for Uber, or search for other important info on the web; knows that I cannot stay in a room with just an electric fan; can take care of me whenever I have allergy attacks; quick to react in cases of emergency; and should know how to take our selfies with sticks, in the future with drones." — **Odette**

* * *

"I mostly travel alone and I team up with like-minded travelers from the hostels or random places. Once a year, I join with my best friends who are living in New York City and we would either go for a road trip just within the United States or we travel overseas. Since we've known each other for a long time, I'm very comfortable with them. Our itineraries are somewhat relaxed, compared to my usual physically-draining activities." — **Jon**

*　　*　　*

"My son during hiking and outdoor trips where we feel like best buddies…" — **Badong**

What have been your most luxurious experiences during your travels?

"Because I am a budget-conscious traveler, I rarely splurge on ostentatious trips abroad. However, the luxurious ones that I did have were mostly press trips organized by a travel brand or a tourism board, in exchange for blog coverage. The first one that comes to mind was a three-week trip across Kerala, India where we were treated to the best that the State could offer, including luxury accommodations all throughout our stay.

While budget travelers will most certainly appreciate Kerala, high-end excursionists who prefer staying at luxury or five-star resorts would definitely enjoy a feast fit for a king! Kerala has fantastic resorts, many of them branded, all over the State. There are resorts by the beach, at the heart of the backwater or in the middle of the pristine forests of the Western Ghats.

Among my favorites were the Vivanta by Taj Kovalam near Trivandrum, the Coconut Lagoon in Kumarakom, WelcomHotel Raviz Kavadu in Kozhikode (Calicut), the Spice Village Resort in Thekkady, Vythiri Resort and Tranquil Resort in Wayanad and the Estuary Island Resort in Poovar. Kerala also houses comfortable five-star business hotels in major cities. I stayed at the Vivanta by Taj Malabar Cochin and the Crown Plaza Kochi, both fantastic hotels in Kochi.

We also had an overnight stay in a luxury houseboat in the backwaters of Kerala. We did this in Alappuzha (Alleppey) around Vembanad Lake, the largest lake in Kerala. Backwaters are where rivers, lakes and estuaries converge with the beach and the sea. It's a charming cultural landscape, especially with the simple rural lifestyle and colorful culture that exists in these backwaters. While we're used to many of these scenes in the Philippines (the village tours might not be as interesting, especially if you come from the rural province), the houseboat stay was a relaxing and pleasant trip through its pristine backwaters.

A favorite part of the trip was the Ayurvedic treatments which we received on a daily basis, Kerala being the center of Ayurveda. We were absolutely pampered while going around, and I constantly joked about becoming a human marinade every time I went for these oil treatments.

Of course, the food was amazing. For centuries, traders have sought Kerala's famed spices. The Romans, Phoenicians, Chinese, Arabs, Jews, and later, the Europeans, all landed in Kerala to trade spices. In fact, Christopher Columbus was headed west to search for Kerala's spices but instead found America. So, it's no surprise that the food in Kerala is an experience worth trying." — **Ivan**

* * *

"Perhaps just like everyone else, the Antarctica Cruise last November 2019 was the most luxurious I've had. Even though I got the cheapest offer for a nine-day trip, the cost was still a huge amount for me." — **Andie**

* * *

"The eight-day luxury cruise my brother and I took in Antarctica right before Covid-19 hit in February 2020. I will forever be grateful to him for taking me along (the most generous birthday gift ever!), as I sure as heck couldn't have afforded it myself!" — **Brian**

* * *

"My husband and I were in Lebanon and were treated by the owner of Chateau Musar, the late millionaire Serge Hochar, to a whole day of wine tasting and food feast in his own private country club. We began in the morning and because we're not used to having breakfast with wine so early, my husband got a little bit of indigestion that almost ruined this splurge." — **Henna**

* * *

"Thankfully, with our travel blogging work, I was able to stay in over 100 luxury hotels around the world. Some of them even provided for our own personal butlers or in-private villas with swimming pools.

I flew business class a few times, a sixteen-day sponsored trip to Antarctica, a private game drive in Kenya, a three-time Caribbean cruise, enjoyed riding a private plane in Venezuela, and was taken to a VIP private tour in Mongolia among others." — **Kach**

* * *

"When in Spain, I stay in a *parador* — a kind of luxury hotel usually located in a converted historic building — usually a monastery or a castle with panoramic views of a monumental city. There are around 90 *paradores* in Spain. I stayed in 87 of them. I have a "gold card" which grants access to breakfast and dinner, an open bar and food served with a variety of *tapas*. I usually take advantage of free cooking lessons when offered.

Parador de Leon, North of Spain, is located on the colossal San Marcos square. In the XVI century, this luxury hotel used to be a San Marcos Monastery and hospital. It has a big chapel inside. Excellent service and superb architecture represented by epic columns, spacious halls, and historical interior that would impress any traveler. I lived like a princess, waiting for my prince.

In the VII century, Parador de Alarcón in Cuenca used to be an Arabian fortress and castle. The hotel combines medieval and modern styles. Its monumental architecture made me feel inspired and safe. Discovering the castle, I imagined myself being part of a historical film because the spirit of the past is carefully preserved. There is even a museum inside, with the interior of the hotel decorated with a collection of majestic paintings. I enjoyed the amazing view, especially when being at the top of its castle battlements.

Somewhere near Paris, The Auberge, is nestled within the Domaine de Chantilly, between the *château* gardens and the Grandes Écuries, where the monumental paintings from the Condé Museum are displayed. Its elegant lines echo a very particular French-style *art de vivre*. Its restaurant, La Table du Connétable, serves inventive traditional cuisine, and the libraries and the open fire in the Winter Garden Bar creates a truly warm and welcoming atmosphere. The spa's subtle lighting and colored mosaics add the finishing touches. It would have cost me an arm and a leg, but I make it a point to avail of the three-days luxury package.

I went skiing at Santa Maria Cap d' Aran, Spain which is one of King Carlos' favorite spots. The valley runs about 25 miles from the French border to the ski resort of Baqueira Beret. The main town, Vielha, and a few dozen small villages dot the valley floor, where stone-and-slate houses are clustered along the banks of the Garonne River and around the spires of medieval churches. Stayed at the Eira Ski Lodge." — **Odette**

*   *   *

"One of my most luxurious experiences was doing a private game drive and staying in luxurious tents with amazing food and impeccable service in the Serengeti National Park in Tanzania." — **Riza**

*   *   *

"Riding the Blue Train in South Africa and Luxury camping in Namibia." — **Luisa**

* * *

"I was exhausted from land travel between Guyana-Suriname-French Guiana, compounded by some issue with my return travel flights. On my last flight home, I chanced upon an advertisement for Caribbean cruises. I have never taken a luxurious cruise on a big ship, so I made a last minute booking for a twelve-day Caribbean cruise with HAL Koningsdam. Fine dining, lots of island tours, and with friendly Filipinos among the crew who took care of me, food and drinks-wise. Probably the only travel where I actually gained significant weight. Other than experiencing minor seasickness, the trip was absolutely luxurious and relaxing at the same time. I'm glad I took this cruise just two months prior to the Covid-19 pandemic." — **Raoul**

* * *

"I went for a flight-seeing tour combined with a crab feast in Ketchikan Alaska. Twice free of charge!" — **Badong**

* * *

"During my visit to Dubai in 2012, I managed to have afternoon tea with my sister at the Burj Al Arab in order to experience what it feels like to be inside the world's only 7-star hotel. The ship's sail exterior appearance is matched by an over-the-top grand interior that

I noticed once I stepped in the lobby on my way to the Sky View Bar through the high-speed elevator, which provided magnificent views of the Gulf. After the visit, a Lexus chauffeur service took us to Burj Khalifa, the tallest building in the world, to continue gallivanting."
— **Dondon**

* * *

"When I availed of a cruise from New York and sailed through Puerto Rico, Caicos Island, and other British Overseas Territories, it was the first time that I had to wear tuxedos every night and endlessly partook of sumptuous dinners accompanied by all kinds of alcoholic beverages." — **Jimmy**

* * *

"Because of my travel hacking, I have been lucky enough to use my mileage points to experience the kind of luxury I would have otherwise not afforded. I've been fortunate enough to have luxurious accommodations in Seychelles, Maldives, Bora-Bora and many other expensive destinations, all for free. The most recent was a stay at the Waldorf Astoria Maldives where paid rates were over US$2,000 per night. I used my Hilton points to stay there for a week, all for free! I've also flown over 100 flights. Other memorable experiences included trying out the best airlines, such as Etihad Airways, Emirates, Singapore Airlines and Qatar Airways A380 — all first-class flights — four times in a row simply by using my mileage points. Finally, I will never forget the Aman Resorts experiences in Bali and other places in Indonesia. Aman takes vacationing to a whole new level!"
— **Rambi**

* * *

"Staying in an overwater cabana in Maldives with my mom. I paid with points and I got upgraded to one of their best accommodations due to my status." — **Jon**

* * *

"I would say the most luxurious travel I had was when my husband and I were in Maldives in 2020, on a relatively small romantic island, just weeks before the Coronavirus pandemic broke out." — **Vhang**

* * *

"I've stayed in some of the best hotels in the world, but by far, my most special experiences were shared with my (then) husband. Our elopement wedding ceremony had me being brought to my husband by boat. The local Fijian men were dressed in their traditional "*sulu*" ceremonial garb. They blew into their "*davui*" — conch shell trumpets — that announced my arrival. When we landed, they carried me over the waters and placed me on the sand where my husband was waiting for me.

During the days leading up to the wedding, we stayed in a *bure*, but that night and for the remainder of our stay, we were upgraded to the main house, which had two floors and two swimming pools. We left this property on a private island in Fiji via prop plane to stay at an equally superb rest house in Malolo Island, where we spent four nights in an overwater bungalow. What a magical sight to see so many fish swimming underneath the glass floor, especially at night!

Ten years later, we celebrated our wedding anniversary at the Maldives, where once again, we stayed in an overwater bungalow. Not only was the water the clearest and best I'd ever swam in, but baby sharks would be seen in one of the restaurants built at the middle of the property.

I couldn't have asked for a better way to mark the beginning and end of my marriage. These properties were the most special, most luxurious places I had ever stayed in, and it was made even more memorable by enjoying them with the love of my life." — **April**

How many photos do you take, and what has been the best photo you've shot?

"As a hobbyist-photographer, I always made sure that I would get my best shots in every photo I took. When I'm not satisfied with that shot, I would take another one. One of the most memorable photos for me was the mirror photo I took while I was having a haircut session in Djenne, Mali. Djenne which used to be one of the most-visited places in West Africa before the rise of Boko Haram aren't getting any visitors nowadays. I was the only tourist when I arrived and I attracted a lot of attention wherever I went. That same evening, I decided to have a haircut. The photo shows the barber cutting my hair with a razor while several curious onlookers (I counted sixteen at one point) were looking with their mouths agape during my haircut session. The interior backdrop was very interesting, light green pastel paint, printed popular African hairstyles, and an assortment of objects arranged rather decrepitly. There was so much story in that one frame." — **Jon**

*   *   *

> *"So much of who we are is where we have been."*
> *– William Langewiesche*

"The best shot I had taken was with the gorillas in Uganda" — **Jimmy**

* * *

"With the advent of electronic cameras and high-tech phones, I tend to take as many photos and then screen them later. The pictures I took of Plaza España in Seville during one noontime summer in 2017 were some of my best photos." — **Raoul**

* * *

"Too many. Before writing books, the pictures told the stories. The date was printed behind some paper pictures using celluloid films. I employed every kind of camera imaginable — Polaroid, instamatic, etc. In 1985 when I traveled to Suzdal, a small monastery town in Russia, I took pictures of the children with their big rainbow colored ribbons. In a few minutes, my Polaroid spun out their pictures. They were in awe: 'Ahhhh Spasiba,' they exclaimed." — **Odette**

* * *

"I take a lot of photos because it's part of our work and you can find most of them on our Instagram account but aside from the main attractions, I love taking photos of local people and sunsets!" — **Kach**

* * *

"Too many pictures! They existed long before computers, digital cameras and mobile phones! The helicopter rides above the Great Barrier Reef, Grand Canyon, Victoria Falls, Ballooning in Kenya, Cappadocia, the Blood trees in Socotra and ruins in Persepolis, Iran were some of the best!" — **Luisa**

* * *

"I take a lot of people shots when my wife and I travel together. When I am alone, I still take a good number of photos, but now of mainly landscapes. I think of myself as having a good eye. A picture of African wildlife in the Serengeti that looks as if it was painted is probably my favorite among all of them." — **Brian**

* * *

"I took a lot of photos every time my husband and I travel, and the very best photo was that of my own city of Lucerne with the Chapel bridge and River Reuss." — **Vhang**

* * *

"I'm mostly interested in nature landscapes and building architecture. As much as I wanted to venture in street photography — as I like also random street walking — I feel the need to respect the privacy of local strangers." — **Andie**

* * *

> *"We travel because we need to, because distance and difference are the secret tonics of creativity. When we get home, home is still the same. But something in our mind has been changed, and that changes everything."*

"My friends know that I take way too many photos when I travel. It is my way to journal my experience. I get criticized sometimes for being too busy taking pictures and told that I'm missing the moment. Well, that is actually my reality and my experience of that moment, and I wouldn't have had it in any other way. I hope some people get that too and stop imposing a point of view that it needs to be a certain way. I took my best photo while my Icelander friend was driving me around the Golden Circle route. During the trip, I spotted a black sheep and a white sheep walking side by side in the middle of the road. It was such a cute sight, and it had an underlying message, so I had to capture that moment on my camera. The picture won in a celebrating diversity theme photography competition in our office." — **Dondon**

* * *

"I'm a photographer but I found that landscape photography doesn't appeal to me in an emotional level. I love portraiture, because it humanizes places. I don't know which among the photos that are my favorites. It's not my call, that's in the eye of the beholder." — **Kit**

C H A P T E R   4
SCARY AND FRUSTRATING STORIES

Did you get sick or hurt in an accident while traveling?

"I was on my last leg of solo travels along the Silk Road in 2019 when I got into a car accident in North Pakistan. My travel group was driving back to Hunza Valley after taking some photos of the China-Pakistan border and some other attractions. I remember having started the trip at 6 A.M. and taking a nap. Around 10 minutes after I dozed off, I was suddenly awakened by a deafening sound with a very painful impact to my chest. The scary part was when we had to drive back for 20 hours after the accident to reach Islamabad so I could take my flight to Thailand and get a temporary medical treatment. The attending physician told me I had a bi-malleolar ankle fracture. I flew back to the U.K. in haste to have my surgery because my (then) husband was there. I was on a wheelchair for two months and underwent a physio-rehab to be able to walk again." — **Kach**

*   *   *

"While in Ecuador, I went biking down one of the tallest volcanoes in the world, Cotopaxi. A few minutes in, at a slope of at least 35 degrees, I hit a dip in the road and fell, hurting myself badly on my left thigh. Apart from not being able to continue the bike ride or even walk upon getting myself up, the wound bruised to the size of a softball and has not fully healed seven years later. I also hit my head hard on the dirt track, but luckily had my helmet on to cushion the blow. My wife, who was biking alongside, just took one glance at me, made sure I was blood-free and continued to whizz down the steep slope fearlessly. I was later picked up by the trail van. I was so traumatized that I have become averse to bike excursions when traveling." — **Brian**

* * *

"I experienced a lot of leg cramps throughout my solo hikes (especially during the first phase of the Everest Basecamp trek from Jinri in Himalayas, Nepal and the Trilogy Day-hike in Cape Town, South Africa). The pain was so excruciating that I thought I wasn't going make it through the end." — **Andie**

* * *

"On the last day of the trip to Iran, the guide suggested to our travel group that we try *kalle-pache,* a Persian dish of boiled cow and sheep parts. The next day was our flight. I began to experience diarrhea and my stomach hurt. At the airport in Tehran, the doctor gave me tablets to take. While I was able to fly from Tehran to Istanbul, but when I reached Turkey, my stomach hurt so bad, I missed my flight to Barcelona. I was fortunate that there was another flight leaving in four hours. After I landed in Spain, I went straight to a hospital in Barcelona." — **Odette**

* * *

"I had severe food poisoning in Jaipur, India. I threw up due to seasickness while crossing the Drake Passage on my Antarctica trip. I fell while traversing a rock gorge in Thailand. I accidentally slipped inside a cenote cave in Mexico, which left me with terrible leg scars. I had the worst mosquito bites (luckily no malaria) in Colombia's Tayrona National Park. I was bitten by a wasp and unwittingly stepped on a large anthill in the Amazon jungle. I had a bad fall while cycling in Zurich. I dropped off a bike into a river while crossing a bamboo bridge in Vang Vieng. I landed in the water on the wrong side of my body from a tree fall. I was shaken by a massive car explosion on the road while I was traveling near the Suez Canal in Egypt. And I witnessed my friend getting bitten by a snake in Australia (luckily, it was not poisonous). Despite these mishaps, I consider myself very fortunate that those were the only things that happened to me during my trips. I had no serious road, sea, or air accidents, no hospital confinement, and, as I write this, I am still in one piece!" — **Dondon**

*   *   *

"While in Leticia in the Colombian Amazon, I figured in a motorcycle accident. I had a double fracture on my wrist and had to fly back to Canada for surgery. I've never been on a motorcycle since!" — **Kit**

*   *   *

"In one of the countless stops while visiting *Estado de Mexico*, my husband and I stayed in a lovely guesthouse that was off the beaten path. Curiously, there were only three guests that evening; and the next day, I discovered why: bed bugs feasted… well, on all of us! All three of us were gone the very next day!

In one of the trips while visiting *Estado de España*, I got so seasick that I caught flu the next day. I was vomiting and literally crawling (from the boat to the rental car to the hotel). A lot of generous bystanders offered help but I refused (I was a mess…dirty and smelly). On my way to the hotel, a police car stopped us. Apparently, we were traversing on the wrong side of the road!" — **Henna**

* * *

"Getting altitude sickness in Tibet. I thought I was going to die." — **Jazz**

Have you had a terrifying experience while traveling?

"This happened in Tegucigalpa, Honduras. I was at a flea market in Valle d'Angeles, walking, laughing, smelling the fresh fruits, haggling prices, buying an embroidered belt. Then all of a sudden, I heard loud screams and gun shots. Total chaos! I ran for cover as the locals screamed, '¡ Sundalos con Pistolas! ¡Corre!' ("Guards shooting, Run!"). Four blocks away, armed guards killed, riddled with bullets three men outside the bus and two men inside the bus! It was horrifying to see the bodies on the sidewalk, people shouting, crying. My heart was pounding! I prayed hard, thank God it did not happen in the market.

Another scary story was at the Gorkh-Terej National Park in Mongolia. I left at 5:00 A.M. to go the airport. It was still dark. Our car had a flat tire, driver had no spare! A Good Samaritan (I thought), named Batu offered a ride. I took my backpack, entered his car. He started a conversation and said we were going for a quick coffee stop. 'No, no we have no time,' I said, but he continued on to a side street. 'I will call the police," I exclaimed but he still drove on. I was so scared and didn't know what to do. Then by some miracle, his car stalled. I quickly ran. A kind Mongolian passerby came to the rescue and quickly hailed a taxi. He then gave a strong warning to Batu not to fool strangers specially tourists and threatened to call the police."
— **Odette**

* * *

"¡Muchos robaron in Cameroon!' — it rhymes and it lingers on your mind. That's what my driver named Juan from Equatorial Guinea blurted when he learned that I am going to Cameroon that evening after the tour. That statement still lingers and I learned it the hard way today. I wasn't literally robbed but I had a scary encounter which shook me up a bit. I was merrily walking around the Boulevard de la Liberti in Akwa, Douala's business district in what was perhaps the only interesting part of the city, on my way to the Cathedral of St.

Pierre and Paul. I was somewhere at the vicinity of Akwa Palace Hotel and I thought it was a nice view with the iconic hotel and the skyline behind it and decided to take out my camera from my backpack and take two photos. Suddenly, I heard some unintelligible shouting but I went on my way, thinking it wasn't my business anyway. The shouting soon grew louder so I stopped in my tracks and that's when I saw a security guard donned in full yellow overalls dashing towards me. He looked angry and was saying in French something about taking photos but I didn't understand fully.

A much older guy wearing a football jersey with combat camouflage shorts soon arrived. I asked what the problem was but he kept on shouting. He pointed at my camera and gestured that he want to see the photos. He wanted to get hold of the camera but I refused and instead deleted the two photos I took in front of him. He was then pointing for me to go to the building on the other side. Sensing it was a trap, I refused, as I pointed to the direction in front of me, indicating I would continue walking. That was when he punched me on the side of my face with such force that I lost my balance. While still struggling to gain my composure with my ears still ringing, I felt someone holding my collar and dragging me towards the building. He shoved me towards the gate but fortunately I regained my footing, else I would have hit hard on it. He indicated for me to go inside but I didn't move knowing that I would lose whatever little advantage I have if not in full view of the public. A crowd quickly formed in front of the building and a few were shouting to the army guy and the guards to let me go. I didn't fully understand what they were saying but I deduced their reactions from the angry rebuttal of the officer. He went inside the gate and came out with an ArmaLite AR-15, which he pointed to the onlookers and shouted for them to leave. When they did, he then turned to me and pointed the gun on my feet, angrily demanding for my documents. I had my passport in my backpack but I only showed the scanned copy on my phone explaining I'm just a tourist.

There was another guy who spoke a little English who did the translation for him. He then took his phone and called someone whom his interpreter said it was the police. A woman who spoke better English came out from the gate and spoke with me as well. I asked what the issue really was and she explained to me that the building where we were seated is a military compound and photos are prohibited. I said I understood and apologized but I didn't take a photo of it but only of the hotel and the street and had it deleted in front of the officer, who wouldn't listen to the woman's interpretation. I told them that there were plenty of photos of that hotel in the internet so I was wondering why everyone can take one yet here I am getting accosted, I added.

One of the security guards was saying something like the police are coming to handcuff me and bring me to the police station. I was already deliberating on the few options available to make an escape. Definitely going to the police station would make the situation even worse. I was thinking of making a dash to the nearby Akwa Palace Hotel just when the officer would go inside with his gun but the petrol station in-between employed the same security agency as the military outpost. When the situation diffused a bit, I began

> *"Traveling is a brutality. It forces you to trust strangers and to lose sight of all that familiar comforts of home and friends. You are constantly off balance. Nothing is yours except the essential things. -air, sleep, dreams, the sea, the sky – all things tending towards the eternal or what we imagine of it."* – Cesare Pavese

my 'Google translation' diplomacy. I asked whether or not we could settle it right there, but he replied, 'you talk with the police because we don't take money here.' That came as a surprise and the thought of dealing with the police in this corrupt country gave me chills. I ran into corrupt policemen all over Africa but only in checkpoints and immigration offices and not in this situation. I was assessing the new territory I've gotten myself into.

A few more minutes had gone by and I made a second attempt to speak with the guard using phone translation. He then brought the officer back who requested for my documents to prove I'm just a tourist. I showed them my visa and entry stamps and the yellow vaccine card for good measure. Upon scanning my passport, perhaps it occurred to him that I was indeed just a tourist after seeing my passport was almost full. He told something to the guard before heading back inside the gate again. The guard typed 30,000 francs or the equivalent of $50 U.S. dollars but I said I only had 10,000 francs. When he went inside the gate to relay the message, I hurriedly emptied my wallet and stuffed the rest of my money at the laptop pocket on the back. I just left 5,000 francs on it. When the guard came back, I showed the upper opening of my backpack with the wallet widely open. He grabbed the 5,000 francs along with the 10,000 on my palm.

The day after that I went to the city of Limbe and had another standoff in the highway when the army wouldn't return my passport for my refusal to give him money. While in Limbe, I was harassed for hours by the sister of a prominent local politician because she didn't want tourists to linger in her city, claiming they have 'a big political problem.' I decided to linger a bit and traveled eastward to Yaoundé, the capital." — **Jon**

* * *

"I had a sad story in Oslo, Norway where I was mugged, rendered unconscious and having woken up in a hospital, stripped of valuables and clothing. Mistakenly assuming that the city was a very safe place, I stepped out one evening to buy a cigarette. The last thing I remembered was seeing multicolored stars before everything turned dark — an obvious consequence of a brass knuckle inflicted upon my face. When I woke up the following day, I realized that I was suffering from a broken tooth, lacerated forehead, devoid of clothing, including the belt with a secret pocket containing my dollars. Missing my eyeglasses, I could not see the people surrounding me, aggravated by my failure to understand Norwegian. I was truly at the nadir of my travel experience.

To make matters worse, I was not able to retrieve my bag which got lost in Amsterdam airport carousel; thus I did not have anything to wear, except for the clothing in my carry-on luggage and the few apparels my wife bought with the limited cash allotment. I was also able to utilize some spare garb from my brother.

Our next tour stop was Communist Russia via Leningrad. The Russian guards with sniffing dogs were astonished with my bedraggled appearance: bandages wrapped around my bruised face and a broken tooth, draped with dirty and disheveled clothing. It would have been my first instinct to be back home, but in view of the huge amount already invested in this trip, there was no point in dropping out from the tour. In hindsight, I wish to thank our Divine Providence, as well as a lady Spanish doctor who kept on tending to my open wounds. When I finally arrived in the Philippines, I discovered that my missing bag was waiting for me inside the house — brought thereto by the KLM people.

Another heartbreaking story was when we were aboard a Trans-Siberian train that took us from Vladivostok to Ulaanbaatar. In one particular instance during the three-day trip, our coach was suddenly broken into by several burly guys and threw us out from our beds, attempting to attack us. Quick thinking saved the day when my stepbrother and I got empty bottles of beer and smashed it unto their faces prompting them to run away." — **Jimmy**

* * *

"It's easy to become cocky and think you know how to protect yourself after years of traveling. When you start to let your guard down, this is usually when you find yourself in a precarious position. As a female traveling solo, I have been in dodgy positions due to carelessness. Luckily, nothing has ever happened, but it could have happened. I've been ripped off by a taxi driver in Buenos Aires, Argentina when he quickly and craftily short-changed me. As frustrating as that experience was, the positive side was that I was safe and unharmed.

You may think it is sexist or double-standard, but as a solo female traveler, I personally like to take extra precaution that a male traveler may not think twice about. I don't go to a bar and drink alone. Does that mean that I may miss out on nightlife or interacting with locals? Absolutely! But my safety is always a priority. For me, drinking alone at a bar creates unwanted attention.

I learned that the hard way in Ko Samui, Thailand when a drunkard tried to follow me to my room. I quickly stopped him by slowly inching closer back to the beach hotel bar, and making sure he sat down again. I also asked the bartender to keep an eye on him. The last thing I needed was for him to know my room number, making an already difficult situation worse.

The last dodgy experience occurred in Isla de Ometepe, Nicaragua and it made me much more cautious. I came back from hiking with my guide friend and started playing cards with a couple in the hotel restaurant. When we sat down, I noticed a group of boys drinking. A few hours later, as we were waiting for dinner, I decided to shower in the shared bathroom upstairs. That hotel had water problems all day and I had to come back downstairs and tell the owner to fix the problem. It didn't take him very long and I quickly went back upstairs. It was already dark, no one was on the floor, and the wind was howling. It was so loud! The bathroom door wasn't fitted correctly as there was a small crack. I was behind the shower curtain and undressed, about to start the water, when two drunk boys from downstairs started peering through the crack and trying to make conversation with me. I was nervous and told them politely in Spanish that I would talk to them when I came back downstairs. When they persisted, I finally yelled angrily, "¡Déjame en paz!" which means, "Leave me alone!"

When I got out, the door was completely wide open. I don't know if it was the boys or if it was the wind, but I was furious! I could have been raped or attacked and no one would have heard me because of that wind! I ran downstairs and told my guide friend what had just happened in imperfect Spanish. When he told the owner, he and the wife just brushed it off, which angered me even more! I threatened to go to another hotel if those guys weren't asked to leave or kicked out. Luckily, the woman of the couple I was playing cards with earlier came to my defense. She understood more English and saw how I was struggling. She spoke to the owners and they kicked the boys out. They weren't even guests at the hotel! I was so thankful for that woman. My guide friend was really young and with my imperfect Spanish, I was not sure if he really understood the severity of the situation.

Since then, I have always stayed in female dorms or rented a room with a private bathroom. Thanks to a female blogger, I now carry a doorstop and a whistle if I am traveling alone. I may never need to use them but they are extra securities for me." — **April**

*   *   *

"I'm grateful that even though I've lived and traveled to some of the most dangerous countries in the world — Iraq, Somaliland, Honduras, Haiti, Venezuela, Papua New Guinea and some other countries in Africa — I've never been attacked or robbed since I usually take precautions and maybe they know I don't have money. Some of the worst experiences though was when I got sexually harassed by some men during my solo travels. We also got some taxi drivers who tried to rip us off and when my (then) husband got pickpocketed in Cusco, Peru." — **Kach**

* * *

"Writing about personal safety reminded me of our Egypt tour in early April 2017. I joined my Korean officemate and his family for a spring break tour to Egypt. Between hotels and train transfers, we were always escorted during handoffs to local tour guides. We got used to that routine. Our main tour guide traveled with us in an unmarked white van, with a designated driver. The last two days of the tour was the weekend visit to Alexandria, during which we were accompanied by a policeman who sat upfront next to the driver. We thought that it was odd to be joined by a police officer at the end of the tour. Our main guide explained to us that the U.S. government specifically requested for police escorts for American travelers to Alexandria. As Asian-Americans, we thought we were "camouflaged" enough to travel incognito. But having a policeman in the front seat was a sure giveaway. During a stop at a roadside oasis, everyone went inside the oasis for toilet and coffee break while I opted to stay

> *"Stripped of your ordinary surroundings, your friends, your daily routines, your refrigerator full of food, your closet full of clothes – with all this taken away, you are forced into direct experience. Such direct experience inevitably makes you aware of who it is that is having the experience. That's not always comfortable, but it is always invigorating." – Michael Crichton*

and sleep in the van. I noticed that the police officer did not leave the van, to watch over me — as part of his orders. I decided to get up and go to the oasis, so he can have a coffee/smoke break as well. We stayed overnight in Alexandria after a whole day tour. The following day, Sunday, we woke up early for a road trip direct to Cairo International Airport, to fly home. That was Palm Sunday.

Hours later at the airport, we heard in the news that there was a suicide bombing at the St. Mark's Cathedral in Alexandria, a Coptic church we visited the day before. I realized that we were lucky to have traveled with a security escort and that our travel guides ensured our safety all the time. Risk is always a factor during travel, and it's a matter of mitigating the risks and scheduling one's tour wisely during safe and peaceful times." — **Raoul**

* * *

"As my travel group were leaving the National Museum of Egypt, I saw a man get shot. Bloodied and lifeless, I saw the man hauled at the back of a van in broad daylight. Startled, I asked my tour guide what we should do and he just shrugged it off like it was just another day." — **Rambi**

* * *

"A man suddenly put his arm around my friend while we were walking in the streets of Serbia. As he would not leave us alone, we rushed into a church to seek refuge and asked the keeper to check if the man was still outside waiting for us to come out." — **Jazz**

* * *

"In Ocho Rios, Jamaica I ended up in a small alley with gangsters where a man approached me and asked for my watch. I didn't hesitate so I gave it away and walked away. It was a cheap one anyway." — **Badong**

* * *

"Yup, I had many unpleasant experiences: from getting robbed and threatened in Krakow, Poland, to being assaulted in Rio de Janeiro. The most horrifying experience of all was being chased by two big men in Port Moresby, Papua New Guinea. I was taking photos of traditional houses in front of the National Parliament on a Sunday afternoon, and there were no other people in sight. My Filipino friend who was in the car started yelling at me to get back in the car because the two locals suddenly showed up from nowhere and started running towards me in a threatening manner. My friend started the car immediately, and I was running towards the car; then, I managed to shoot myself inside a running car just as I kicked one of the guys, who managed to almost got hold of my feet. It was like an action movie!

Unfortunately, random opportunistic violence was rampant in that city in 2014, so I could only think of the unimaginable if I had been caught. In yet another episode, in 2019, I was detained in Qatar's Doha central police station overnight in a cell with other prisoners because they suspected I was a spy when police saw me taking photos in front of the Amiri Diwan government complex earlier in the night. Fortunately, I was released in the morning as they realized that I was just an unsuspecting tourist taking pictures. One lesson I learned is to be very careful when taking photos of government buildings because it can land you in jail!" — **Dondon**

* * *

"In Africa, my survival instincts were on full throttle. I took calculated risks in deciding at times to jump off the truck and leave my travel group. Sometimes I'd jump off with a couple of my travel buddies, other times I'd go solo. My most memorable experiences came out of these independently-planned excursions. In the Democratic Republic of Congo, we walked through the streets of Kinshasa (formerly Zaire), regarded as one of the most dangerous cities in Africa, because we wanted to experience the city for ourselves. In Mali, we ran through the chaotic streets of Bamako, just a few short weeks after the bloody siege of a hostage situation at a hotel where twenty-seven innocent victims had been killed. In Gabon, we rode a night bus to Libreville, with a driver who seemingly had a death wish, given the suicidal speeds that he drove our vehicle. In Nigeria, we explored the crowded markets of Lagos, always on guard for presence of Boko Haram, even though we had no way of identifying who they were. In Sudan, on my own and in search of food, I explored the streets of the harsh desert towns, unwittingly becoming the day's source of entertainment for bemused locals. In Guinea, on another solo venture, I ran into villages while trying to constantly remain on the lookout for people that looked sick, for fear of getting infected with the Ebola virus.

Thankfully, I survived all these perilous voyages. I had, at all times, carried a police-grade pepper spray and a personal alarm with me, but never once did I feel I needed to use them. The first-hand experiences and knowledge I gained made it all worthwhile for me. Undoubtedly, my survival instinct kicked in and was at its highest ever in my life during these times. Taking risks was all part of my travels. I'd like to have my own perspectives about the different places in the world and the different environment that people live in. Life is not a bowl of cherries for most people, but then the dangers were sometimes overstated and sensationalized by the mainstream media. Not all of Nigeria is terrorized by Boko Haram. Not all of Sudan is in a warzone. Travel is not all wine and roses to those who truly want to experience the world.

Yes, I was occasionally sexually-harassed. The behavior of the Nigerian immigration officer who flirted with me and threatened to keep my passport so he could keep me in his quarters frightened the pants off me. The border patrol officer who entered our truck and demanded a photo taken with me, calling me his future wife, alarmed my travel group. These kinds of behavior are unheard of in other parts of the world.

The young man at the backpacker's lodge in Zimbabwe who wouldn't take no for answer when he wanted to take me out on a date tested my patience. Taking what I thought was a kind offer from a campsite guard in Morocco for a motorcycle ride back to my tent and finding myself being taken for a long ride outside the camp without my permission was a terrifying experience. These were just some examples of my encounters, but thankfully nothing led to a sexual assault or anything more serious than that. I'm afraid it's all part of the travel experience." — **Riza**

What was your worst experience with transport, customs or border control?

"One of my worst experiences occurred largely because of my ignorance. I traveled to Bulgaria with the notion that I could obtain a visa upon arrival. But when I reached Sofia, immigration officials said that I could not avail of visa on arrival as they required a Schengen Visa — a short-stay visa allowing its holder to circulate around the Schengen area comprising of 26 countries in Europe — which would take at least fifteen days to process. Instead of gaining entry to Bulgaria, I was promptly arrested like a convicted criminal. I was quickly transported to Hungary where I was, again, detained until travel administrators transported me to Montenegro, where I was incarcerated anew. It was outright humiliating as my hands were always locked in handcuffs throughout the ordeal, with a sniffing dog in my constant presence.

I was not allowed to stay in hotels as I would be entering territorial space without travel authority. Thus, I spent my supposed sleeping time in non-territorial spaces or "no man's land," which was literally the area where passengers retrieved their luggage from the baggage carousel. Being seen thereto, devoid of proper and hygienic appearance, I could not even avail of the toilets without being accompanied by guards. At that moment, I knew exactly how a convicted felon felt like; it was clearly the most humiliating experience in all my journeys.

While being detained, I thought of possible ways of talking my way out of the situation. I considered blustering about my contacts in several local newspapers and Philippine government officials. But upon deeper reflection, I decided that it would not have been a wise move, as I would be besmirching their reputations due to my grossly negligent act of not being able to secure a Schengen visa.

I was very fortunate to have an attractive lady companion with a pleasing personality who sweet-talked an immigration officer who finally agreed to let me go in exchange for a financial sum. When I was finally released, I suddenly found beauty in freedom, including the hideous faces of the security personnel who arrested me. This harrowing experience took all of five (5) days." — **Jimmy**

* * *

"Attempting to smuggle Delimondo corned beef back to Canada and always having it confiscated." — **Kit**

* * *

"I prefer traveling by land if time is not an issue and I have several crazy backbreaking journeys that I kind of regretted doing while I was at it, but which I look back now with fond memories. I remember an incident from my Guinea-Bissau to Guinea-Conakry border when I crossed it in January of 2019, albeit a painful one. I managed to apply and collect my Guinea visa only on January 2 at Bissau. Embassies were closed for a few days on holidays and I was already getting bored with the city. After collecting my passport with visa around noon, I rushed to a shared taxi terminal and got a lift to the border town of Contabane. Flying around West Africa was so expensive; so rather than shelling out US$400 for a sixty-minute flight, I chose to toughen it out by traveling by land. I got myself a motorbike to bring me across the border to the city of Boke where buses and shared taxis were available onward to Conakry.

The road from Contabane to Boke was not a straightforward path. It was more of a trail than a road. We had to go through three security checkpoints from the Guinea-Bissau side then another six in the Guinea area. In two of those points, guards asked for a bribe which I was prepared to give. A sentry remarked that I was the first Filipino to cross that God-forsaken border route, and gave a parting shot with the words, 'Philippines good country!' I couldn't stop rolling my eyes. My motorbike driver didn't speak a single word of English and I knew very little French so it was a hilarious conversation. The route wasn't scenic landscape-wise but we passed through villages where communities were still living in mudbrick houses made out of loam, mud, and sand and covered in thatched straw roofs, passed through half-a-dozen creeks where I saw men, women, and kids in various states of undress. At one point, we had to cross a huge river and had to get on a *pirogue* — a native dugout canoe. We arrived in Boke quite late and pushed for the capital city of Conakry. It was a very distressing ride as I had to share a dilapidated four-seater car with seven other passengers. I regretted not paying for the vacant two seats at the front to have a more comfortable ride.

To make matters worse, the back of the seat had a protruding steel bar so I had to crouch throughout the entire journey. Getting into the center of Conakry was another challenge because of the numerous checkpoints; I had to show my passport every time. Some of our fellow passengers didn't have proper documentation so they had to bribe the officers at every checkpoint. My calvary didn't end there: because the taxi wasn't allowed entry in downtown, I had to hire another motorbike to finally get me to my hotel. I finally arrived in Conakry at 2 A.M. after an exhausting thirteen-hour journey on various modes of transportation — a shared-taxi, motorbike, *pirogue*, another shared-taxi and another motorbike — crossing thirteen security check points, and plenty of body aches and joint pains."
— **Jon**

* * *

"I've had three 'unfortunate' immigration experiences — the first one was at the Djibouti-Somaliland border where the immigration officers were very rude and wouldn't allow me to exit the country for Hargeisa. But after a few hours of waiting it out, they let me go just the same.

The second incident was in Riyadh, Saudi Arabia. I was traveling around March 2020 when the Covid-19 restrictions began and I was already on the flight from Istanbul to Riyadh, when they announced that Filipinos who were initially allowed to get visa-on-arrival weren't allowed to enter the country. I stayed at the airport for more than thirteen hours. Immigration officials blamed the airline personnel for allowing me to board the flight even though I wasn't allowed to enter the country. I remember the Saudi immigration officials were very rude.

The third episode was at the U.S. Immigration. There was an instance where I was stopped at the checkpoint and had me undergo three different screenings upon entering U.S. territory. The officers were nice but I've been interrogated in three separate instances." — **Kach**

* * *

"Crossing the border to Mauritania where I had to wait six interminable hours because some corrupt officer in West Africa wouldn't let me through without a bribe." — **Luisa**

* * *

"In February of 2020, I traveled by local van from Eswatini to Maputo, Mozambique. I applied for visa upon arrival, but I was held up for interrogation as the immigration officer questioned my country of birth (Philippines) and being a U.S. passport holder. He called another office, inquiring if my passport or citizenship was valid, etc., which I figured out from the little bit of Portuguese I understood. I was not sure if he was "soliciting" a bribe, but I knew that I was qualified for visa on arrival at any Mozambique border crossing. I was sweating bullets, but I simply kept my cool. He asked inappropriate and somewhat racist questions, perhaps to rattle my cage. Perhaps he was incredulous that a non-Caucasian could somehow travel with a U.S. passport.

I did not waver and stayed resolute with my request to be granted a visa upon arrival. Meanwhile, I was worried that the van would leave without me. Fortunately, I had befriended some of my co-passengers on the trip and they were aware that I was a foreign tourist traveling locally. They had to wait almost an hour for me. They were finally relieved that I was able to rejoin them on the bus." — **Raoul**

* * *

"The first thing that comes to mind was the experience that my wife and I had in Entebbe, Uganda. We had a layover and wanted to explore the city but the immigration officers were wary of a young couple traveling from country-to-country and only spending a few days in each. They probably thought we were smugglers and instead of letting us out, held us in a small room and told us to wait for the next flight out of the country." — **Rambi**

"In the Republic of Congo, I learned to navigate through Immigration and its corrupt officers who would brazenly ask for a 'donation' in addition to the visa fee. I'd endure a standoff with these officers so as not to pay a single penny." — **Riza**

* * *

"Driving from Democratic Republic of Congo to Angola with police (showing off their rifles) as our 'guards' instead of guides (and a driver) and negotiating for how much. I say it's the best US$300 pay we've given. They did the mediation at the border immigration (I could not remember seeing any structure but the presence of armed police from somewhere). We got back safe and checked off that list of visiting a new country. I hated the experience of going through immigration in the Congo (Democratic Republic of Congo and Republic of the Congo), Burkina Faso, etc." — **Henna**

* * *

"With a valid travel permit from the representative office of Kurdistan in the pocket, my husband and I approached the Turkish/Kurdish border post in Silopi. There we were told by the Kurdish immigration that Filipino passports need a proper visa from Baghdad, and we had no choice than making a u-turn back to Turkey. I still have not been to Iraq." — **Vhang**

* * *

"It was in Tel-Aviv Airport both inbound and outbound airport immigration, I think this was normal for all solo backpackers. Airport Immigration Officers were very strict if you came from Arab

> *"A passport shows government officials who you are, where you were born, and how you look when photographed unflatteringly." – Lemony Snicket*

countries as a tourist rather than as a worker. My work visas in KSA and Kuwait were ignored during interrogation. Instead, they scrutinized my Egypt, Lebanon and UAE tourist visa stamps." — **Andie**

* * *

"In Turkmenistan, despite having a letter of invitation and a round trip ticket, I had to wait at the airport counter for three-and-a-half hours while they contacted the Consulate." — **Odette**

* * *

"My worst land border crossing was during a trip from Bolivia to Argentina. I had to catch a cramped van from Tupiza before 4 A.M then endure a 2-hour journey. Upon reaching the border town of Villazon at 6 A.M. amid the freezing cold, I was greeted with a massively chaotic sea of people waiting for their passports to get stamped out of Bolivia and then stamped in Argentina. I crossed the border around midday after what seemed like an eternity waiting in the queue. Then a long bus ride awaited me, as I traveled from La Quicaca to Salta, Argentina, finally arriving in the evening. That was undoubtedly a very stressful travel day!" — **Dondon**

* * *

"When I was about eight or nine, my family and I were on vacation in Switzerland and decided to drive to Liechtenstein for the day. As we were close to the Austrian border and my mother had successfully applied for an Austrian visa, she asked my Dad if we could stop by the border, for her to cross over to Austria (just for the sake of using the visa — those caught with the travel bug understand) and back. She decided to bring me with her. While we were crossing into Austria, the Austrian immigration officers questioned whether my Mom had the legal right to bring a minor (me!) into Austria or if she was 'kidnapping' me. This accusation infuriated my Mom, and we were forced to wait in limbo for almost an hour as they phoned Vienna to verify. When all was cleared, my Mom and I walked into Austria to buy a postcard from a kiosk and walked right back into Liechtenstein to join the rest of the family. Seeing my Mom so livid (as she is normally very mild-mannered) made this episode unforgettable." — **Brian**

* * *

"I once almost got deported from a country before I got stamped in. I flew from Jeddah, Saudi Arabia to Djibouti, and on my flight there, I decided to wear a local outfit that I had bought in a local market in Hargeisa, Somaliland a few years prior. It was a purple abaya with an attached hijab. The nun who was sitting next to me gave me a smile of approval.

The female customs officer also approved of my look. The interaction seemed to be going well until she asked for my boarding pass from the flight. Fine. She then looked in my passport for the Saudi Arabia stamps. When she asked where it was, I pulled out my other U.S. passport.

I have two U.S. passports. One was valid for 10 years, and the other was valid for 4 years and had different passport number. At that time, I had to send out both passports to their respective embassies in Washington D.C. because I needed a business visa for Saudi Arabia and I needed a tourist visa for Djibouti. I wasn't sure if I was going to receive the Saudi Arabia visa in time for me to send that passport to the Djibouti embassy before the start of my trip. When the customs officer asked where my exit stamp for Saudi Arabia was, I naively volunteered my other passport, not knowing how much trouble this would cause me. She looked confused that I had two valid passports and asked more questions about my stay. She then took both the passports to show to her manager. Meanwhile, I stood at her station, waiting patiently for fifteen minutes, still not realizing something was wrong. She led me to her boss, who asked me about my local outfit. I told him that I wanted to respect the local culture which is why I wore it. He asked me if I was Muslim to which I replied, "No." He looked frustrated and said, "Get your bag and go upstairs." So, I did.

Once I went upstairs, I realized it was the boarding area of the airport. It started to dawn on me that I may be in trouble, but I kept my cool. No one had told me anything. I thought the officers wanted me upstairs so they could check my bag.

Thirty minutes later, an Ethiopian Airlines employee came upstairs and told me the officers wanted me to leave. "Holy shit," I thought, "I'm getting deported without even entering the country!"

I started crying and asked him why. I told him I had already been to 20+ African countries and I had never had an issue in any of those countries.

He told me that I was suspicious because I had two passports and didn't have a plan while in Djibouti. I'm not sure where the miscommunication laid, but I told him I had a three-day tour set up the next day, but I was going to couch-surf for the first time ever that night. I was really agitated, and cried harder. As he walked away, he suggested that I start figuring out which flight they could put me on and if I had a contact in Djibouti, to contact him immediately.

I took his advice and called the tour operator immediately. What an angel. Ken was about to sit down for lunch, but he immediately came to the airport when he heard me crying on the phone. I told him I was getting deported and I didn't know why or what I was supposed to do. A free lunch and a couple of 'Narcos' episodes later, I was released. Thankfully, my time in Djibouti was less eventful, and I knew to never again use two different passports in the same trip!" — **April**

Describe your most uncomfortable transport ordeal.

"Flight from Honolulu to American Samoa Pago Pago International airport took approximately five hours. After two-and-a-half hours, the captain announced, 'The weather semantics in the cockpit is not working.' In other words, he cannot see if it was too cloudy. We had to return to Honolulu. Passengers looked at each other in disbelief! I had questions in my mind,

'We have to fly back another two-and-a-half hours, with or without weather determining factor?' 'Is it safe?' Thank God, we reached Honolulu safely. We stayed overnight with free accommodations and were able to fly after two days.

Another uncomfortable transport ordeal was at the Chatham Islands airport. I had to check-in early because the wind was the determining factor if flights would take off that day. Flights go for only six months of the year. So, we were finally able to fly after six hours of waiting. One of the passengers declared that he would sit in the cockpit and have a conversation with the pilot! There was no door to the cockpit, only a black curtain. And this was allowed? When I asked the flight attendant, she said 'it's ok'. After 10 minutes, I reminded the flight attendant again to tell the pilot, I was nervous because the pilot might get distracted! The plane was stuffed with boxes of ice crates and cray fish. In case of an emergency, the passengers would have to jump over the crates to the nearest exit. We were only fourteen passengers. I told the flight attendant 'This is it! I am going to write a complaint to Air Chatham.' Only then did the talkative passenger return to his seat. We landed safely at Wellington, New Zealand." — **Odette**

* * *

"My travel companions and I were on our way to a Weta workshop for 'The Lord of the Rings' Studio in New Zealand. We didn't realize that we had to pay a fare in NZ$ and we only carried US$. The bus fare was only about NZ$5. When we handed the driver US$50, he wouldn't return the change back and replied that he only got New Zealand dollars. Fortunately, an old woman helped us out settle with the driver." — **Badong**

* * *

"Our most unpleasant ride would have to be in Tanzania where our small shared taxi was suddenly loaded with eleven passengers in a seating capacity of five." — **Vhang**

* * *

"I've traveled mostly overland through Central and South America, Southeast Asia and even some parts of Central Asia and Eastern Africa by hitchhiking, local buses and even by boat, which were mostly uncomfortable." — **Kach**

* * *

"I have a deep-seated fear of flying in airplanes. However, during my 21 months world trip, I've probably done over fifty flights and flew in some of the smallest aircraft, and landed in some of the scariest airports, like in Lukla at the Himalayas. A few months before I arrived at this mountain-top airport, two planes crashed. Imagine what I must have felt when I saw the pilot reading a newspaper while we were mid-air at some point during our flight at the Himalayas. That almost shocked me to death!

I got left behind by airplanes and trains a number of times due to my running very late, although sometimes it wasn't even my fault, like the time when the Ryanair check-in counter closed out on me since it was already one hour before the flight needs to depart (though it should have only been 50 minutes!). I got left behind for some tours that I've previously booked, such as the one for Mont-Saint-Michel in France, where I was 15 minutes late for the coach. Finally, I missed a 4 A.M. sunset tour in Masada, Israel, because I woke up at 6 A.M.! I learned to be always on time to avoid stressful situations like these." — **Dondon**

*　*　*

"I'd never thought I'd get affected by a train strike in France. I always hear about these strikes in the news. But since I am thousands of miles away from home, it never really bothered me. In 2010, I unfortunately got stuck in Hendaye, the first train station after the border between Spain and France, on the way to Bordeaux, because of a train strike.

Everything was going as planned. I left Madrid Chamartin at 10:30 P.M. and got off at Valladolid Campo Grande at 1:20 A.M. to change trains. Then I boarded another coach headed for Hendaye at 1:54 A.M. It all worked like clockwork. Trains left on time and arrived on time.

We were scheduled to arrive in Hendaye at 7:10 A.M. where I was to change trains for Bordeaux. I was a bit groggy but I noticed we stopped moving at about 7 A.M. at Irun, the last station of Spain.

By 7:10 A.M., passengers were getting restless. Obviously, we all had trains to catch in Hendaye. Then came in one of the train staff announcing that the train could move no further and would not be able to proceed to Hendaye because of a train strike in France!

We were advised to get off and take a taxi to Hendaye which was nearby. The taxi cost was €11,50 which I got to share with three other passengers. We crossed the invisible border, so invisible in fact that I did not realize we were already in France until the driver pointed to the Hendaye Station. Unfortunately, there were no buses. I had this wild idea since I saw an Avis car rental across the street. But there were no available cars. So, I was stuck in Hendaye, no train to Bordeaux, hoping to get on board a 2:10 P.M. provisional TGV to Paris that would pick up all the stranded passengers along the way.

The same thing happened again a year later. In 2011, I had a Eurail pass with me. And when I arrived at a train station in Paris, we found out there was a strike. Rather than wasting the entire trip, this time we were successfully able to rent a car and drove through Europe visiting France, Switzerland, Liechtenstein, Italy, Vatican City, San Marino, Slovenia, Austria, Hungary, Slovakia and the Czech Republic, thanks to the train strike." — **Ivan**

Describe the most stressful situation you've encountered while traveling.

"I flew from Istanbul to Djibouti in October 2018 with an e-visa in hand. As a Philippine passport holder, I could either get a visa on arrival or an e-visa. And since I was arriving quite early, I decided to get the latter so that I wouldn't have to wait long in case the visa on arrival officers would not be available that early in the airport.

We promptly arrived at 5 A.M. and I was one of the first on the queue. When I handed over my e-visa to the immigration officer, he looked over my documents for a long time then asked me what my purpose for visiting their country was. He went inside to an office and told me to wait on the seat behind the queuing area. I was surprised that no one attended to me even after everyone already went through the immigration procedure. I spoke with the officer when it was apparent, they'll leave me cold waiting and he advised me to just wait.

After two hours a guard asked me to move to the departure waiting room which is on the next floor above the arrivals. The solitary cafeteria started opening by then and I ordered a coffee to fight the fatigue that was slowly sinking in. Three hours into the waiting time, I went down the stairs that had access to the immigration area for outgoing passengers. I stubbornly picketed at the stairs to attract their attention. Realizing that I won't go away easily, they started making calls and promised that they'd have an immigration officer come within an hour. He eventually arrived 1.5 hours later and I was sort of grilled for the purpose of my visit. I kept asking why they had to detain me when I have all the necessary documents while the rest was allowed out easily. I didn't get any answer. After five hours I was out and on my way to my hotel." — **Jon**

* * *

"I was denied entry due to the absence of a visa. Another exciting experience was when I tried to seek admission to Bermuda. This time I was being deprived entry because of having too many visas from different countries in my passport thinking that I could be a spy, a criminal or a person who would cause harm to the community. After showing him all my papers, one documentation that saved the day was my business card, which to the authorities was indicative that I was a normal and law-abiding person. I also discovered that the citizens in Bermuda were people whose ascendants were from West Africa who arrived thereto as slaves. I was shocked, however, to see at their Museum the graphic presentation of their present-day social ranking where the Bermuda people were placed on top of the social levels and at the bottom were the Filipino domestic helpers." — **Jimmy**

*   *   *

"The disarray, panic and disruption during the viral pandemic had caused a lot of stress. I was on the road traveling full-time when the pandemic happened. When borders started shutting down and Manila declared a lockdown, I found myself in a situation where I had nowhere to go. Luckily, my ex-husband had offered me shelter and I was able to ride it out for a couple of months at his home in Delaware, U.S.A." — **Riza**

*   *   *

"My flight out of Manila for Canada via Japan and the USA (March 2020) at the peak of the Covid-19 fear and hysteria…" — **Kit**

*   *   *

> *"All journeys have secret destinations of which the traveler is unaware." - Martin Buber*

"Being stuck in Riyadh, Saudi Arabia. I was allowed to board the flight but the government announced a rule while we were on the plane that Philippine citizens were not allowed to the country because of Coronavirus and it was on March 1, even though we came from Europe and I live in Europe (not Manila) ... We originally had trips to Oman and Yemen that we ended up cancelling." — **Kach**

* * *

"August 2018, I was supposed to fly from Belgrade to Luxembourg via Warsaw. LOT Air suddenly cancelled the flight. I was in a long queue at the customer service with the other irate passengers, but I was running out of time. So, I called their New York office to arrange for alternate flight to Luxembourg. I was routed through Milan on two different airlines that did not have courtesy bag forwarding, with only half-an-hour for connection.

So, I had to claim my luggage at Milan and check-in again for the flight to Lux. With five minutes to spare, my luggage came out of baggage carousel — I grabbed it, ran as fast as I could up to departure area, and even pleaded with security personnel to allow me jump the line. Fortunately, they were very accommodating, gave me directions and off I ran! The check-in counter was already closed, but the ground crew was still there — luckily! They checked me in quickly and put me on special vehicle on the ground to rush me to the aircraft. Phew! It was a close call. I only had three days to visit Luxembourg and I did not want to miss that trip — because I wanted to complete all the countries in Europe — and that nearly got derailed due to LOT Air's poor fleet management that summer." — **Raoul**

* * *

"In Iraq, entering a museum and some of the heritage sites were a pain; even if we had the necessary supporting papers. The soldiers or guards would be calling somebody on the phone and let us wait for very long hours inside the car... only to tell us to come back the following day." — **Luisa**

*   *   *

"When I was in college, I enrolled in a Peace and Conflict Resolution Program because it included a trip to Israel, Egypt and Palestine to study the origins and development of the Israeli-Palestinian conflict. It was on one of our jaunts to the West Bank, during what came to be known as the Second Intifada, that our bus sped through a section where Israeli soldiers were to our left armed with rubber bullets, while on our right lay Palestinian teenagers ready with throwing stones. A skirmish was definitely about to begin. That got the entire class' andrenaline rushing!" — **Brian**

*   *   *

"As we left Brazil, we had a short layover in Peru before heading to Toronto, Canada. As we deplaned, I realized that I left my Bose canceling headphones on the plane. I asked airport staff and they told me the only way to get it was to head out of the secured area and seek permission from the airline staff. This was around midnight and most crew were already home.

Simultaneously, my wife found out that she needed a Canadian visa even though she is a U.S. immigrant. She needed to secure the visa before they would let her board the plane. All this we had to accomplish within the next 90 minutes. Like some miracle, we ended up getting my headphones back and the Canadian visa on time."
— **Rambi**

* * *

"Lining up for a visa in rural Bolivia across a long queue in a dilapidated street crisscrossed by pedestrians and cargo and passenger trucks under the heat of the noonday sun was one. Taking twelve hours sitting on a bus traveling from Chitwan to Pokhara in Nepal on a mountain road with no toilet facilities on the bus or on the roadside was another. Note that I had to answer the call of nature in-between our bus and a deep ravine below." — **Jazz**

* * *

"In Kinshasa, I had a 'private tour' with a guide I met on the street (which was perhaps my mistake). Like a typical tourist, I took a photo of a church building because that was the only structure that can pass for a 'beautiful and intact' edifice in the entire city. Lo and behold, a 'policeman' suddenly sprang out of nowhere and told us that we should pay him for taking the photo. He claimed it was prohibited and that we were violating the law. I suggested erasing it from my camera. The next thing I knew, he was already grabbing my camera. In a matter of seconds, I was surrounded by a sizeable mob. I think I walked inside and realized later on that I was already on church grounds because I saw a man emerging from the building garbed in a priestly vestment. I thought he would be our 'savior' negotiator but it turned out he was rooting for the crowd. I was afraid we'll be mauled so I had to give in and hand over my camera to the 'policeman.' Since my new camera was secured in my backpack, losing this old one wouldn't hurt. The next thing I knew, I was running for dear life. It could have been much worse. In hindsight, that camera was useless anyway, since I had the accessories with me and that battery was running low." — **Henna**

* * *

"I had the pleasure of getting into an argument with a local in Tiebele, Burkina Faso after he accused me of taking a photo of him when I really took a photo of a sign (and not realizing it, he was in the photo too). He was so aggressive that I passive-aggressively flipped him the bird as I went back to reading my book in the car. This incensed him and he grabbed the car keys and called the police. After the locals talked to the police officer, we had to go to the police station where the chief held a "court hearing" in which both sides presented their case. I can't speak French so I had to trust what was being said by the guides, and after hearing both sides, the judge kicked me out of Tiebele." — **April**

* * *

"My most stressful event was at the airport in Rome for my flight to Tel-Aviv. I remember rushing to Immigration and to boarding gates. I forgot that Israel flight boarding is already strict on the port of departure. Another forgettable moment was when I missed my flight from New Delhi to Kuwait due to messed-up train schedule. For that, I had to purchase a new one-way ticket which was almost triple the amount of a regular flight. My third unpleasant encounter was at the Ashgabat airport immigration, where there was a problem in my Transit Visa and flight discrepancy as both of them were secured from Dubai. Finally, the experience of being a chance passenger stressed me out. Since I was an aviation employee, I have this benefit of availing flights at a cheaper cost, provided that I needed to wait to close the check-in counter first before they can hand-over my boarding pass."
— **Andie**

* * *

"In Halifax, Canada where I only had a limited number of hours to get back to the ship, I went for a short hike and got lost. I found my way back following the sound of cars passing by but I almost missed the ship. They were already calling my name on the public address system..." — **Badong**

* * *

"From Maputo, Mozambique, I took the public bus. I went to the station early to buy a ticket. It turned out that the station opened at 7:00 A.M., which in African time was still 8:00 A.M. But then we had to wait for the bus to fill up. I had to leave my small luggage in the U-Haul cart. I refused to put it in the back, for fear it might get lost with all the luggage, furniture, live chickens, and pets. Finally, after two hours, the bus was filled. After driving for an hour, the bus suddenly stalled. We had to wait under the scorching sun for another hour until a rescue bus arrived." — **Odette**

* * *

"Traveling from Rwanda to the coast in Tanzania (a bus ride of eighteen hours through very dry and remote areas), our rather primitive bus broke down in Singida. In the middle of the night, we had to sleep on our seats with all the mosquitoes in the air, hoping that the vehicle would be repaired; but then, the driver announced that this was the final stop, and we were kicked out. We had to look for a place to stay. As this was not enough, I was experiencing a stomach problem, probably from a fish soup I ate along the way, so we had to change our program and head to Arusha hospital for a check-up." — **Vhang**

Were there any cultural practices that pushed your buttons?

"In some African, Asian and Middle Eastern countries, people still practice Female Genital Mutilation (FGM). They believe that the young girl becomes clean and beautiful after the removal of the body parts that were considered unclean and unfeminine." — **Luisa**

* * *

"Driving through the Mopti dessert to reach Timbuktu, we stopped at a circumcision rite being performed in the Bandiagarra cliffs, Mali — a UNESCO World Heritage site. In the village of Songho, the circumcision grotto was a place where young boys undergo this rite of passage. The young boys spent their time playing songs and reviewing the paintings on the rocks. They would stay here for twenty-one days and nights. During the circumcision proper, every woman should leave the village. We were not allowed to see the rites. We stayed from afar. Each boy went in an open area covered by curtains. Then I heard loud cries." — **Odette**

* * *

"Yes, it is the Islamic practice of seeing men hanging out in public places (i.e. coffee shops, parks) and women were nowhere to be found." — **Riza**

* * *

"I consider myself to be a very tolerant and respectful traveler. That said, on occasion, the noisiness of mainland Chinese tourists, the snootiness of local Parisians who look down at foreigners when we try to speak in French, the blatantly corrupt practices of Italian cops, and the annoying habit of Koreans to spit in public places, have bothered me." — **Brian**

* * *

"It's not a cultural practice but a general perception from a lot of places. Being a tourist, there were places where people treat me like I can pick money off from trees. I wouldn't generalize but it happens in a lot of developing countries. I'm just a simple tourist who travels light, but I get annoyed when random people try to befriend me and ask for money in the end, or ask for my shoes, phone, headphones, etc." — **Jon**

* * *

"Some people, like myself, can get initially confronted and anxious about having to be fully naked in the Finnish sauna, Russian banya, and Japanese onsen with a lot of others around." — **Dondon**

* * *

"It was a visit to the Karen tribe, Thailand's "Long Neck Ladies" near Chiang Mai in Thailand. Long Necks' villages became an attraction, like a human zoo, where for 300 Baht you can take a photo with women dressed in colourful outfits, with a neck so long that her head looks almost like a poppy flower; and how these girls can manage to grow up wearing these heavy metal rings every day just for tourists, and still give a perfect smile to its many visitors." — **Vhang**

* * *

"Women wearing the *niqab*, or Islamic face veil, which fully covered their faces with only eyes and nose showing in Qom, Iran. Women not able to use the swimming pool and other facilities in our hotel in Tehran, bothered me…" — **Jazz**

TRIVIA: According to Human Dignity Trust, as of August 2021, there are 71 countries where lesbian, gay, bisexual, and transgender people are criminalized.

There are 11 countries where LGBT activities may be punishable by death: Afghanistan, Brunei, Iran, Mauritania, Nigeria, Pakistan, Qatar, Saudi Arabia, Somalia, United Arab Emirates, and Yemen.

CHAPTER 5
FUNNY AND BIZARRE STORIES

What's the funniest experience you've had while traveling?

"During a trip to Luang Prabang in Laos, I stayed in an individual room at a hostel with my then-girlfriend and another friend. As the 'experienced' traveler, I suggested that we leave our passports in the room instead of carrying it throughout town. We all found our own hidey holes. I had the great idea of putting my passport in the room fridge's freezer compartment, as it was empty. So, I wrapped it in a plastic bag and left it in there. A few hours later, when we returned from our day out, I checked on my passport and saw that it was now wet and all the passport stamps (which is a traveler's prized possession) were smudged! It turned out that while we were gone, a brownout occurred and the freezer began defrosting...this 'experienced' traveler sure ate some humble pie that day!" — **Brian**

*   *   *

"Missing flights because I drank too much at the airport out of boredom. Happens a lot." — **Kit**

*   *   *

"I attended a gathering with the world's most-traveled people at the invitation of the President of the Micronation of Liberland where he also stamped my passport in front of the TV crews, and we all tried to land on the soil of 'Liberland' by chartered boat from Serbia. But unfortunately, we were chased by the Croatian coastguard and they did not like the idea of us, including the President, to set foot in his own country of Liberland. So we were just in Liberland's territorial waters of the Danube river. On the way home, I could not enter Croatia anymore, my passport having an official red stamp of Liberland. But we had to return our rental car in Croatia! So, I had to stay in Novi Sad, Serbia for one night while my husband returned the car; then proceeded to Belgrade where my husband later joined me to stay on the river house boats, before flying back home." — **Vhang**

* * *

"I once left my passport on the seat pocket of the plane. I was already queuing for immigration when I realized I didn't have my passport with me. I alerted the airport officials who immediately had to chase the plane which was on its way to the hangar. What an embarrassment!" — **Jon**

* * *

"It happened when my stepbrother and I availed of a public toilet in Lhasang, Tibet due to an urgent urge to avail of them from overeating. The toilet papers which we just wiped from our behinds flew to our faces due to the strong winds inside the toilet." — **Jimmy**

* * *

"I always make it a point to purchase a souvenir item whenever I travel. One time, in Mumbai, India, I was approached by a balloon vendor who showed me a few samples of huge balloons that he was selling; for some eerie reason, he convinced me to buy a whole set. But when I opened the pack, I discovered that the balloons were fake. I tried to chase the vendor, but couldn't find him anymore.

I discovered later on thru the Internet that this type of racket of scamming customers by showing big balloons, but selling tiny ones were very prevalent." — **Vhang**

* * *

"Climbing Mount Makiling during the rainy season when there were hundreds of thousands of *limatiks* (blood leeches) with fellow traveler Jon turned out to be a harrowing experience. We were totally unprepared for the ruthless attacks of the *limatiks* and I ended up screaming my way through our climb to the summit. I can laugh about it now, but at that time, I was absolutely petrified when a *limatik* got into my eye and started sucking blood through my eyeball. Thankfully, Jon was quick and resourceful to get the *limatik* out and saved my eye." — **Riza**

* * *

"I showed up on the wrong date for a flight, arriving a day early, and even lined up at the check-in counter. I held up the line, as it took a while for the ground crew to realize my actual flight was scheduled for the next day so they could not find my name on the passenger list. I was so embarrassed, to say the least." — **Raoul**

* * *

"As I usually book a lot of airline tickets for my family, I've had my share of unforgivable blunders. The most memorable one was when I booked my mom an airline ticket for our family vacation for the following year. A year later, I did a cursory confirmation of our tickets and to my horror, discovered that the ticket I booked was for the same date the previous year!" — **Rambi**

* * *

"I lost track of time wandering around the Chatsworth House in England. I didn't realize it was already closing time, so I got locked on the garden grounds. I had to jump off the wall to get out. There was no more public transport from there, so I had to walk my way to the nearest town carrying my large backpack. On my way there, I got chased by a large herd of cows in the English countryside!" — **Dondon**

Describe a "lost in translation" moment that you've had in your travels.

"I was at a mom-and-pop store in Hong Kong shopping for souvenirs. I paid with a crisp new 50HKG dollar bill. The shop owner perused it and held it up against the light to inspect it. The assistant (his son) was getting antsy with his dad's fussiness. The shop owner then gave me a 20HKG dollar bill as change, and I did the same thing — held it up against the light, pretending to inspect it the same way he did with my original bill. He wasn't amused, but the other customers laughed, including the owner's son. Unfamiliar with the Hong Kong currency, I would have absolutely no idea if the bills were fake or not.

Undeniably, part of it was probably the storeowner "profiling" me. (He did not peruse other customers' HKG bills prior to my turn). Call this a 'lost in currency exchange' moment. I guess the joke was on the storeowner." — **Raoul**

*   *   *

"The most memorable moment happened during my winter visit in the Republic of Georgia. I had three hitchhikes on a chilly night where all the drivers knew absolutely zero English (first one was when I got lost at night in Tbilisi, the second one was in Gudauri, and the third instance was when I was going back to Tbilisi)." — **Andie**

*   *   *

"On my last night in Tokyo, Japan, I decided to go to a karaoke bar. I thought it would be like the bars in North America, where one could sit and watch others sing karaoke. As I approached the desk, a language barrier ensued as I tried to explain what I wanted to do. The employee kept telling me the price for an hour and the price for a drink. I finally agreed. We walked out of the elevator, and the employee opened a door. To my horror, the room was empty. In the States, there is one common room where people sing karaoke. This was not the case in Asia: karaoke rooms were in abundance! I belatedly realized that I had rented my own private room for myself for an hour! "Oh, what to do, what to do? Ahhhhhh, f*ck it," I thought. I sang my little heart out. I had enough practice in the

Philippines for the past two weeks. As the last minutes of my hour wound down, I started singing 'You Give Love a Bad Name' by Bon Jovi. As I sang 'Shot through the Heart', there was a bang on the door. "Yes?" I asked, annoyed that I was disturbed and yet afraid that the person would tell me my voice was too loud and too irritating. "Ten minutes", he barked. "Oh ok, thanks", I said as I continued singing. How embarrassing!" — **April**

* * *

"In the remotest provinces in China, not one word of English was spoken, so I usually had to deduce the list from the menu — which were in Chinese — and chose at random. One item on the menu turned out to be snake soup." — **Vhang**

* * *

"To cross the Mediterranean Sea, I took a ferry from Algeciras, Spain to Tangier (the armpit of Morocco). While walking to the hostel, I met two Japanese backpackers who were staying at the same place. During check-in, we saw a flyer for a public bath combined with a Moroccan massage, what they called a *hammam*. After a long journey, it sounded like a relaxing first cultural experience to have in Morocco. Oh boy! Well, we found the *hammam*, entered and communicated by way of hand gestures (as no English was spoken) that we would like to have a treatment. They handed each of us a small bag which, to our surprise, contained disposable underwear. We hesitantly changed and were

led to a very hot room where we were instructed to lie down on our stomachs. Suddenly, a loin-clothed old man popped out and asked us if we wanted '*hammam*' (this was the extent of his English), to which we nervously nodded. He proceeded to bring out a sponge and black soap. He jumped on the back of the first Japanese guy and began pummeling him hard with the sponge, which I found hilarious at the time. After a few seconds, he went to the second Japanese guy and did the same thing. I was downright amused. Finally, he got to me, jumping very roughly on my back (his feet thrashed my hips) and dug this sharp sponge that felt like sandpaper all over my shoulders and back. Talk about painful! To make matters worse, the old man began heckling while doing this, going from one person to the other, then to the other, bringing on such intense pain (presumably to remove dead skin cells, although I was pretty sure he removed a good share of live skin cells) for what seemed like forever. Definitely not the soothing bath and massage that we had imagined! Lost in translation, indeed!"
— **Brian**

* * *

"Commuting by shared taxi in most major cities of Central Africa is a very competitive sport! It was especially so in French-speaking countries where I struggled mightily. Taxis never had fixed routes but they normally took passengers at any place close to the main road, depending on their bids. Catching an empty taxi was always fortuitous because, most often, the driver will take you straight to your destination as his main route.

Basically, you stand on the roadside, wave on the oncoming vehicle and when the taxi stops, you have a few seconds to shout your destination and how much are you are willing to pay, say for example, '*Nlonglak, pres de l'ambassade de Chad, cinq cent!*'. Try it, super-fast. It's tongue-twister, no? If the driver agrees, he will honk; but if does not, he will just leave you biting the dust with nary a word. Obviously, since I couldn't pronounce the names of the places properly, I get left behind a lot. Soon, I learned to stand at a curve or the common taxi stops so I would have more than the usual time to shout out my spiel properly. It was fun and frustrating at the same time.

If you want to get a specific destination, you can outbid anyone. In one instance, for example, I was already running out of time heading to the airport in Libreville because I was quoting normal fares and no one will take me. I tripled my bid and the driver stopped to the irritation of most passengers because the taxi will have to take a long detour to my destination. But when I'm not in rush, I don't really mind the many detours; in fact, and I take it as an opportunity to see the neighborhoods." — **Jon**

*   *   *

"At St. Petersburg, we asked a cab driver to take us to a park or tourist place and we ended up in a strip club." — **Badong**

Did you have some weird or surreal experiences?

"Yes, many, due to ingested substances." — **Kit**

*   *   *

"A few surreal experiences come to mind. The Kekak Dance experience in Bali. Meeting Maasai Warriors in Kenya. Seeing and smelling *guman* bodies being burned in the rivers of Nepal. And being hedged by a school of eagle rays in Bora Bora, French Polynesia." — **Rambi**

* * *

"While in Bolivia, my wife and I went to a witch doctor's house to have our futures read through coca leaves in a cup. Apart from how creepy the shack looked (with dead, dehydrated animals hanging on its walls), and being served donkey milk in a shot glass, my wife was eerily told we'd have five kids. To this writing, it hasn't happened…yet (well, there is still time though. Haha!)." — **Brian**

* * *

"The Druids' pagan ceremony celebrating the Mabon, during the autumn equinox in September 2014 at Stonehenge. I had to wake up early dawn to get picked up from London for the two-hour drive to the Stonehenge, in time for the sunrise ceremony. It was all a blur now, given my fatigue at the time, but the smell of weed and exotic incense will forever be etched in the olfactory part of my brain." — **Raoul**

* * *

"At Chichen Itza, there is a place in the center where, when you clap with cupped hands, you hear the echo of a bird which is indigenous to the area.

Another surreal experience I remembered was when I saw people lined up to enter the biggest pyramid in Giza so I decided to go in too. My friend refused to go with me, so I decided I was going with the people whom I saw in line before me. I just got a ticket and followed inside. I walked inside the tunnel trying to catch up with the group who was in front of me. All I saw was a long tunnel with no one in it. Panicked, I called out to the people but no one responded. With my heart pounding and my legs feeling heavier and heavier, I ran out of that tunnel befuddled: "Did I really see people go in or was it the desert playing tricks on my eyes?" To this day, I still couldn't explain how that incident happened." — **Jazz**

*   *   *

"I visited the Equatorial Monument in Ecuador, the marker for the equatorial line determined by the French in the 1970s. A few decades ago, a more accurate measurement of the equatorial line was identified some 250 meters off the monument. It was fascinating being on the equator. My friends and I even checked the GPS on our phones. It was fun experimenting with gravity along the equatorial line, it being less than the opposite poles -- an egg can steadily balance on a nail, water drains in opposite directions depending on which side of the line you are at, and the most amazing of all, the resistance test where arms above my head with interlocked fingers can be easily pulled down which was much harder to do on either side of the line. Looks surreal but intriguing science." — **Dondon**

*   *   *

"In Namibia, it was fabulous to see the Milky Way shining over the sand dunes of the Sossusvlei desert.

A bizarre incident happened when I was waiting for an early flight to Paro, Bhutan. I stayed overnight in Calcutta Airport Hotel, and while I was sleeping, some stranger just entered the room and slept on the adjacent bed." — **Luisa**

* * *

"I'm Roman Catholic. During one of my solo travels in Central Asia, my mom messaged me and asked if I would be going to Mashhad in Iran. I was not familiar with that Holy City. I later found out that the first city I would be visiting after crossing the border from Turkmenistan to Iran was Mashhad. My mom then instructed me to visit the Imam Reza Shrine and make a wish because miracles do happen there. Customarily, one could only enter the shrine if he's a Muslim but luckily, I was able to get in, touch the shrine and say my prayers. Let's just say a miracle happened after that." — **Kach**

* * *

"During a hiking expedition in Tarak Ridge at Mariveles, Bataan, I was left walking alone as my companions had gone ahead. I had this weird feeling that someone was following me as I heard footsteps. Instead of looking back to see who it was I, instead, ran double time until I caught up with my friends who told me they had a similar spooky encounter." — **Badong**

* * *

"I arrived in Freetown Sierra Leone Lungi International Airport at 2:00 A.M. It was a dark and rainy night. I was preparing to hail a water taxi when all of a sudden, porter boys grabbed my single luggage and ran fast shouting in Creole. I initially panicked, thinking that they were stealing my only luggage. It turns out that they were simply tugging to get an Express ferry ticket and earn a small commission. What a relief!

I paid the money. Put on my life jacket and we sailed off. Upon reaching the mainland, my friend's brother who was supposed to pick me up was not at the airport. What to do? A guy with tattoo on both arms offered a ride in his van. I was afraid to take the taxi alone at 3:00 A.M. So, I hopped into this stranger's van. He introduced his brother-in-law, who was curiously in a formal suit and tie. I closed my eyes and prayed hard, hoping this was not a set-up.

During the trip, I started a conversation. I said I am from California, have traveled to 261 countries, and an author of five books. I handed him my customized book mark. 'Really? Congratulations!' he exclaimed in utter amazement. He shook my hand, asked me to sign an autograph. 'I have not sat this close to a writer,' he

> *"To awaken quite alone in a strange town is one of the pleasantest sensations in the world." – Freya Stark*

exclaimed. He gave his name: Charbel Raad. 'My driver will bring me home first,' he uttered. I discovered to my amazement that he lived in a mansion with seven cars.

'Wow! You're rich!' I bellowed. 'My driver will bring you to your hotel. He will pick you up tomorrow — I will show you around.' It turns out that he was from Lebanon who was assigned to Freetown, Sierra Leone, chief of logistics of a diamond mining company. He did show me around and introduced me to his friends. 'Be careful what you say, she's a writer,' he joked. He made sure I was safe, texting me everywhere until I exited the airport because he was well-aware that airport personnel can gave tourists a hard time. At the day of my departure, Charbel even called Teresa, the security officer at the airport exit to whizz me through. It was a surreal experience with a pleasant ending." — **Odette**

*   *   *

"One weekend in the summer of 2012 when I was still living in Italy, I decided to venture north to Valle D'Aosta. The Alpine valley is one of the most beautiful parts of the country located just below Monte Bianco (Mont Blanc), the tallest mountain in Europe. I was traveling alone that time because my colleagues opted to stay in Turin so I have to find ways to entertain myself after the usual touristy stuff during the daytime.

"I asked the hotel personnel if they could recommend a place that would be nice to hang out and meet people in the evening. They told me there's a party happening a few blocks behind the town in the woods. After dinner, I decided to walk towards the venue and shortly after, I saw the entrance where I had to pay to get in. The ladies manning the entrance told me that the actual party site is in middle of the woods and motioned me to follow a trail. The pathway was lighted with light bulbs. I was jittery and excited, not knowing what I'm getting myself into. A few minutes later, I saw some more lights emerging from the shadows, which I soon found out was coming from a bonfire. The trees that surrounded the bonfire, which was were decorated with colored light bulbs creating a magical effect on the entire place. It wasn't only the lights that made it look surreal but the partygoers who were wearing psychedelic costumes and fairytale-looking dresses, as well. I soon realized that I stumbled upon one of those clandestine 420-friendly hippie parties. I forgot what took place that entire evening; but I could only remember waking up in my own hotel room the following day." — **Jon**

*   *   *

"I went to Burning Man at the Black Rock Desert, Nevada in 2017 and on one of the last days of the festival, my new friend asked me to join him in riding our bicycles out to the Temple. The Temple was exactly what I thought it was: it's holy, it's welcoming, and it had an incredible energy inside and surrounding it. But, hell, it can be intimidating and overwhelming as well, if one's not ready. That day, I wasn't ready. I wasn't ready to see the massive amount of posters dedicated to loved ones lost. I wasn't ready to hear the chants reverberating throughout the Temple. I felt unprepared because I hadn't known what the Temple was before I arrived on "the *playa*", and hadn't prepared anything to dedicate to my deceased mother.

Immersed in guilt, I slowly walked around the grounds, and while reading the posters, a small postcard stuck on a bench caught my eye. The postcard read: 'Forgive others, not because they deserve forgiveness, but because you deserve peace.' I was struck by that statement, and let the words roll inside my mind and on my tongue. Before I had time to allow myself to fall into deeper thoughts, my friend found me outside, and we rode off to see another art piece.

Two days later, I was saying my goodbyes to my friend, and he asked me if I had pulled a card out of his deck yet. "No," I replied. He then pulled out a pack which looked similar to tarot cards, and fanned them out in his hands. I pulled one out and turned it over. It read: "Forgiveness". I couldn't believe it! I started bawling as I hugged him. Clearly, it was a sign that I needed to develop this virtue.

A week at Burning Man had depleted me of all my energy, and made me aware that it was incredibly special. I would never have another week like this again, and I was certain I would not return to Burning Man for a second consecutive year. Since work had denied the vacation request a year ago, I decided not to buy a ticket to Burning Man. As the months got closer to the event, I realized how much I missed the *playa* and how it was the only place in the world where I felt truly free and loved without judgement and labels. It was also the only place in the world where I never had to take a billion photos and where I allowed myself to truly live in the moment, creating and reinvigorating deep connections (it helped that there was no cell phone reception).

About two months before the event, I saw a Facebook announcement stating Burning Man was selling an additional thousand tickets to anyone who would to take a Burner Express Bus ride to the event. I decided to try my luck on the day of the sale, and wouldn't you know it? I got one ticket! I left it all up to the universe to obtain a ticket, and she provided!

Knowing my first Burn was of epic proportions, I went into the second one more subdued. In my first year, veteran participants warned me the second Burn was the worst because of high expectations. I had no idea why I was meant to go to Burning Man again this year, but I knew I had to spend more time at the Temple. Consciously, I knew that my mother's death anniversary would be significant that particular year as it was the ninth year and she would finally be in heaven. I wrote a tribute to her and to a few others lost to untimely deaths. I wrote two gratitude tributes and taped them up on some free wood as well. Tears were shed. I moved on.

On Monday after sunrise at Robot Heart, two new friends and I were biking toward the opposite end of the *playa* when I saw a copious assortment of country flags. I made the guys stop and we spotted a man rolling some flags and placing them in his car. He looked at me and asked, 'Are you a princess, a goddess, or a dream?' With a smile, I said, 'Whatever you want me to be.' I asked him about the flags, and he told us about his goal of creating a prayer drum circle in every country in the world. At that time, only four countries had prayer drum circles.

"One fine day, it will be your turn. You will leave homes, cities and countries to pursue grander ambitions. You will leave friends, lovers and possibilities for the chance to roam the world and make deeper connections. You will defy your fear of change, hold your head high and do what you once thought was unthinkable: walk away. And it will be scary. At first. But what I hope you'll find in the end is that in leaving, you don't just find love, adventure or freedom. More than anything, you find you." – Isa Garcia

I looked around and was in complete awe. Everyone who knows me knows that this is my mission in life: to visit all 193 member-countries of the United Nations. It was a lofty goal I set when I was still in my twenties, and it was only a few years ago when I seriously started to pursue this goal. It was one thing to know it as a number; it was another to see it physically represented. I almost started crying. I was surrounded by almost all of the flags of the countries of the world, and here, on the *playa*, in the middle of a made-up city, was where I accepted my destiny: I will visit all 193 countries." — **April**

CHAPTER 6

INSPIRING STORIES

Is there any place in the world that you wouldn't dare go?

"None. No such place for me." — **Riza**

*   *   *

"I am not afraid to go anywhere, for as long as there's a way to go there." — **Luisa**

*   *   *

"Not really, I would like to climb up on the list. I should not be afraid of going anywhere in this world." — **Vhang**

*   *   *

"None, but there are places that I already visited, which I would never return again, like Djibouti." — **Kach**

*   *   *

"No. I visited Iran in April 2013, and that trip made me more curious about the surrounding areas, places that are not normally on a traveler's radar: the 'Stans, Abkhazia, and Dagestan. The more I travel to remote or unconventional places, the more I realize that I have so much more I want to see!" — **April**

*   *   *

"Nothing I think. If I have the opportunity, time and budget, I would love to visit them all. I even had a chance to join a competition for a free trip to space in Axe Apollo Space Academy (a once-in-a-lifetime opportunity to fly on Earth's Middle East Grand Finale with an all-expense paid experience)…" — **Andie**

* * *

"Nah, as long it's already visited by humans…" — **Badong**

* * *

"No! While there are places that I don't dream about visiting, I would go anywhere for sure!" — **Rambi**

* * *

"I will go anywhere where I will not die in the process!" — **Jazz**

* * *

"I won't push my luck on regions that are controlled by extremists, like the junctions of Chad, Nigeria, and Burkina Faso. Libya is very appealing with its rich history but I won't venture in that area without guides. Yemen has always been at the top of my list but won't go anytime soon while a civil war is raging." — **Jon**

* * *

"I would not know as there's a propensity on my part despite my age and other limitations that I would still dare to go…" — **Jimmy**

* * *

"I would never say 'never' with finality, but I am somewhat tentative about visiting countries with civil unrest (such as those in Middle East) or those countries in Africa that can be visited via border crossings mainly." — **Raoul**

*   *   *

"No, but I'd prioritize visiting the safer parts of war-torn countries instead of the more conflict-affected areas." — **Brian**

*   *   *

"No..." — **Kit**

Describe the best day you've ever had during your travels.

"Celebrating my birthday sightseeing in Luxor in the daytime, then going back to our river boat for dinner with the staff singing a birthday song for me, then unwinding in the pool drinking Dom Perignon while cruising the Nile River and capping it with the best sex I've had in my life." — **Jazz**

*   *   *

"Undoubtedly the best day of my travels was when I proposed to my now-wife at sunset at a tree house in the middle of the deep jungle. That was during an overnight stop after an exhausting zip line full day adventure at the Gibbon Experience in Laos. I was so scared I would drop the engagement ring while swinging from kilometer to kilometer of zip lines before reaching the tree house!" — **Brian**

*   *   *

"In the Sistine Chapel at the Vatican, when my husband proposed to marry me." — **Vhang**

* * *

"I cried when we arrived in Antarctica in December of 2016. I never thought that I would be able to do it at 28 years old and on our honeymoon. My (then) husband and I got married in July 2016, then I got this amazing news of going on a sixteen-day Patagonia and Antarctica trip which was fully-sponsored by a Hurtigruten company. We were the first travel bloggers they hosted on this complimentary trip." — **Kach**

* * *

"Among the many best days that happened during our visit to Seychelles, I was able to finally buy the long sought after *'coco de mere'* as part of the decorative item in my place of residence." — **Jimmy**

* * *

"There were so many unforgettable moments, but one recent experience that comes to mind was the day I arrived in Santiago de Compostela in Galicia, Northern Spain, after weeks of walking in the Camino de Santiago. I've been traveling all my life, but I never really saw myself as a 'pilgrim', and the idea of shifting my mindset to one of pilgrimage was something I'd always wanted to do. Even though I was raised a Catholic, I was never a spiritual or religious person. It was exhausting both physically and mentally, and many times I wanted to give up.

The Camino was the perfect metaphor for life in general. You just keep going. It also gave me the opportunity to slow down (my life was normally one of constant movement) and just be alone with my thoughts, having shared a part of me with strangers who later became good friends. Finally, there's something about walking… the gradual wearing down of our bodies with every step that is symbolic of our time here on earth. It was both morbid and rejuvenating at the same time. I arrived in Santiago on the morning of my 41st birthday, and I wept. It was truly one of the happiest days of my life. I would do it again in a heartbeat." — **Kit**

* * *

"After spending more than three weeks touring the GCC (Gulf Cooperation Council) countries in the Middle East, I met up with my relative who was on a work assignment in Tel Aviv. After a hectic schedule crisscrossing between Oman and Kuwait, I planned some downtime to relax. We stayed in a hotel right across the Bugrashov beach. As this was my second visit to Israel, I decided to take it easy — the first few days in Tel Aviv were mostly spent relaxing on the beach, reading a book and basically 'chillaxing'. I never thought I needed some serious down time after a hectic travel schedule, but relaxing on the beach was just 'what the doctor ordered'!" — **Raoul**

* * *

"My best day would have to be waking up before sunrise, going for a swim, and having breakfast by the beach. This would be followed by a nap, then go for some scuba diving or sightseeing. At night, we would start with the best local cuisine we could find, followed by a sporting event or a show. We would then finish the night off with drinks and hookah. Pure bliss." — **Rambi**

* * *

"I was very happy to be able to go to Afghanistan and Socotra-Yemen." — **Luisa**

* * *

"One of my best travel days was on 26th September 2013 in Monrovia, Liberia at the 17th PCL Compound, a big military compound with seventy personnel. I was met by twenty-two military personnel complete with a band singing our Philippine National Anthem. There was a giant billboard sign: 'Welcome World traveler: ODETTE AQUITANIA RICASA has traveled to 247 countries.'

They all asked 'Why are you in Liberia?' and 'Why do you travel solo?' I was assigned a bodyguard as we went around the city in United Nations white-colored four-wheel drive vehicle. We drove through Broad Street, Clay Street, and McDonald Street for the post office. We went to the mosque, an orphanage, and a refuge Baptist church. Unforgettable!" — **Odette**

Describe the most interesting people you have met while traveling.

"I visited the Hiroshima Peace Memorial and Museum as part of a special tour with the JICA-sponsored Japan-Philippines Friendship Program. We had the opportunity to meet Akihiro Takahashi, who survived the atomic bomb on August 6, 1945. He described that fateful day down to the last gruesome detail. After surviving the initial blast, he started walking towards his home — while still dazed and disoriented. As he was about to cross the river, a fireball ensued. He jumped into the water because of the intense heat. The water helped neutralize the searing heat, which he said saved his life. He showed us his blackened fingers, his damaged hand and deformed ear. He had to undergo therapy and hospitalization for more than a year. He was in and out of hospitals for surgeries and therapy. At times when he felt despondent over his health situation, he said he had to remind himself of his mission: to share his experiences so that future generations may be reminded of the horrors of war and nuclear weapons. He even became the director of the Peace Museum. Takahashi-san continued sharing his stories and experiences until 2011 when he died at the age of 80 — it was a life well-lived, considering all his health issues as a Hiroshima survivor." — **Raoul**

*   *   *

"I met a young French and American couple who were traveling on a bike from the hostel I was staying in Hama, Syria in 2009. That year, they were traversing across the Mediterranean beginning from France, across the Balkans, covering the Middle Eastern countries that were at the juncture of the sea, eventually cutting across Northern Africa all the way to Morocco, and passing over to Spain. It was already their second expedition. The first one was a year ago when they cycled all the way from Scotland to the Philippines (after flying from Singapore to Manila). They sold all their possessions, resigned from their respective jobs to able to pursue their dream of traveling the world. I hung out with them on evenings, learning about their journey which has tremendously inspired me to discover the world myself." — **Jon**

* * *

"Each and every person I have met has a story and it has always been fascinating. People come into your life for a reason, and you can learn so much from them, even for short while. Most people tend to stay away from countries with active war zones, but what I discovered is that if the locals are comfortable enough with you, they will open up and share their stories with you. And my goodness, their stories will always be so much more fascinating than your own. I can think of not one, but two extremely interesting people I have met while traveling.

When I was in Herat, Afghanistan, I was introduced to a man named Abdullah. He is Russian, but now considers himself an Afghan. He came to Afghanistan as a KGB soldier and was captured. After his capture, no one from the Soviet Union came to rescue him. Abdullah was very angry at his country, thinking they had abandoned him. After some time, he decided to convert to Islam. Later, when Herat was occupied by the Taliban, Abdullah joined them. Now, he is a guide for the Jihad Museum in Herat, and has an Afghan wife and family.

> *"We find ourselves after airplane doors close and wheels touch the heavens. We discover the maps to our hearts when we lose the maps to this world. Wander, and find home in the people you meet. Wander, and find home inside yourself."* – Tyler Knott Gregson

Obviously, I was impressed by his story. Imagine, this man having been part of two globally-formidable military units, stood in front of me, and yet, Abdullah was polite and friendly. As the only woman in my tour group, he could have easily dismissed me, but he shook my hand. He acknowledged me. What a pleasant surprise to see a man who was dressed in Afghan clothes, but who still held on to his Russian roots, and who forwent his adopted country's practices by offering a small but significant gesture to a woman.

Another interesting person whom I had the pleasure of meeting was a *babushka* woman in her 70s named Eugenia. When she was in her early 40s, she and her mother were evacuated from the Chernobyl exclusion zone. After staying in a 'refugee camp' for a year, she was allowed to come back to her home because she was unmarried, never had children, and was over the age of 40. When I met her, she lived alone with her dog and raised livestock. She was very self-sufficient: she had to cut wood for fire every day. She didn't have radio or television so she didn't know much about the news in Ukraine or in the world. She only learned of them when her neighbors would drop by and tell her. Eugenia was so strong, so self-sufficient, so self-determined. It was hard not to admire her." — **April**

*   *   *

"One of the most interesting persons I have met was a young, selfless gentleman by the name of Sahr who wanted to build a school for his village in Sierra Leone. Sahr was the son of the Chief of Bondorfulluhun town in North East Sierra Leone, an area known for its diamond mines. Sahr was given a piece of land by his family and wanted to use it to build a school for the children of Bondorfulluhun and its nearby villages. The residents of the villages in the Kondo district were refugees who were displaced from their homes during the Sierra Leone Civil War." — **Riza**

* * *

"Omar Sharif, reciting the poem with fireworks, light and sound show at Pyramids in Egypt in one of the special occasions that I attended. It was so magical!" — **Luisa**

* * *

"I've met a lot of interesting people in my travels but the one that stuck out was Per whom I met in a Travel Century Club meeting in Brussels, Belgium. He had visited all 193 countries and spoke over ten languages! He went around the room to talk to us in our native tongue. He even spoke to me in Tagalog! Amazing!" — **Rambi**

* * *

"There are countless highlights of my trips, one is meeting very interesting yet the humblest of human beings. My husband and I had the chance to spend a whole day with the late Serge Hochar, famed owner/proprietor of Chateau Musar, a winery in Lebanon. He treated us like top notch cherished friends." — **Henna**

* * *

"From all the high calibre travelers I have met, it was really hard to choose who among them stood out as the most interesting, as they were all very compelling in different ways. But if I were to name three, I would pick Harry Mitsidis (the world's most-traveled person) from Greece; Don Parrish (veteran American adventurer) from the USA and Michael Runkel (a world-famous travel photographer) from Germany." — **Vhang**

What was your most memorable local experience in your travels and why?

"One of my most memorable local experiences was the Wagah-Attari Border Ceremony, a daily military practices that the security forces of India and Pakistan have jointly followed since 1959. It was an awesome encounter beginning with immigration, when we literally bid "bye-bye, India" as we pushed our luggage through customs; then said, "hello, Pakistan!" with just a few steps.

The Wagah Border Ceremony did not disappoint. The sheer magnitude of enthusiasm which Indians demonstrated was beyond my expectations. The blustering parade by soldiers from both sides were simply electric! We were very fortunate to sit in the front row of the grandstands. There were thousands upon thousands of heads in a sea of people. There was a coach who encouraged the Indian crowd to cheer. A Hindustani battle cry, '*Hindustan Zindabad*' echoed in space, followed by prolonged boisterous cheers.

The Pakistani counterparts also tried to arouse the Pakistanis to acclaim. Many others were invited to dance in festivity as patriotic marches filled the air. A company of Gujaratis performed *Garba* (a traditional dance rendered around a lighted clay lantern) together with Punjabis dancing along Bhangra music. We did not shy away from being part of the celebration. We lent our shouts to the already-electrified atmosphere! The BSF (Border Security Force) soldiers stood in attention awaiting instructions.

As the BSF soldiers exchanged performances, the crowd kept chanting patriotic slogans. By then, the number of Pakistanis had now grown significantly and their every roar was reciprocated by an even louder holler from the Indian group. The atmosphere was electrifying and intense! I surely got goosebumps! The scene was truly indicative of India's unity in diversity. The drill was characterized by elaborate and rapid dancelike maneuvers, which have been described as 'colorful and blustering.'

Another memorable local experience for me was in Tarawa, Kiribati. I met locals Elizabeth and Emmy who worked at the

New Zealand bank across my hotel, the Boutique Tarawa. Most people walked barefoot, even going as far as the airport without sandals or shoes. Aren't we lucky we have more than three pairs of shoes?

Emmy is married to a Filipino seaman, and loves to watch Filipino movies. 'We will pick you up tomorrow to show you around the island,' she said. 'Tomorrow is my birthday,' I replied. 'Oh…Happy Birthday — we will order a cake for you,' she acknowledged in return.

On the day of my birthday, I was picked up very early and was taken to Beti Island, Bairiki, Biketawa. I attended church services at the Sacred Heart Cathedral while sitting on the floor, as there were no pews in church. Houses were built on stilts; there were no sidewalks except for a two-lane highway. We enjoyed a fried chicken and rice meal with raw fish for lunch and I drank *Karewe* (a Kiribati palm sap wine). Atio strummed his guitar, amid boisterous laughter, exchanged high-fives, and enthusiastically sang 'Happy Birthday' in Gilbertese language.

I also visited Nauru, the least-visited island in the world, because of its strict visa entry requirements. The island is only sixteen kilometers long. Despite being surrounded by the Pacific Ocean, the island lacked places to swim. I managed to dip in the waters somehow at the Community Boat Harbor, close to the Menen Hotel, where I stayed.

During my stay, I asked a waitress at Ruby's Fast Food if there were any Filipinos here. She pointed the next door Capelle and Partner

Company. Can you imagine a big Filipino group working at a foreign company? It had a supermarket, a bakery, a Digicel office. They worked in IT, web design, accounting departments. We posed for selfies and group pics. All of a sudden, I became their Tita. It was fun interviewing them. Some have stayed over sixteen years. When I gave my bookmark, everyone wanted my autograph. Vicky, Danuba, Lai and Michelle showed me the accounting office, the furniture store, and the dress makers shop." — **Odette**

* * *

"My memorable local experiences was visiting a small village of Himbas, the last semi-nomadic tribe of Namibia, We had a unique opportunity to meet and learn by listening to their stories of life and understanding their beliefs. A great heartfelt, once-in-a-lifetime adventure!" — **Vhang**

* * *

"Few people realize that North Korea or the DPRK, as they prefer to be called, is not as isolated as we perceive it to be. There are regular tours that leave from Beijing which almost anyone can join. It was an exciting prospect that I finally got to do in 2014.

I made inquiries and bookings several months in advance. The tour company in Beijing took care of the visa application process. You get issued a Tourist Card for your visa where they place entry and exit stamps. And unfortunately, you don't get to keep it.

The cheaper option to get in was to take a 24-hour train from Beijing to Pyongyang via Dandong. But for those who couldn't stand long-distance travel, there were flights from Beijing to Pyongyang. Since taking a train allowed me to see views of rural North Korea, and it was the cheaper option, I chose the coach. The train left Beijing Train Station at 5:27 P.M. and arrived in Dandong, the border city of China, on 7:17 A.M. the following day.

In Dandong, there was a nearly three-hour stopover, enough time to walk around the city to see the Yalu River Bridge, which was bombed by the U.S. Air Force during the Korean War, and the Sino-Korean Friendship Bridge which connects Dandong with Sinuiju, North Korea on the other side of the river.

The train for Pyongyang left at 10:00 A.M., but we had to go through China Immigration first before leaving. Once everyone was on board, the train departed for Sinuiju just across the bridge. Looking outside on the right side of the train while crossing the bridge, I saw the end of the broken Yalu River Bridge.

The trip across the bridge was just ten minutes. But we arrived at 11:10 A.M. because of the one-hour time difference. From there, there was a long wait inside the train since Immigration and Customs procedures were performed on board.

Our Immigration Forms were collected. The officer then asked us to present all our mobile phones for inspection since they jotted down the brands in our Immigration Form. Then the Customs Declarations were collected. The officers went from one cabin to another searching every bag to make sure prohibited items were not brought in. They were very strict about Global Position System (GPS) trackers. If a passenger's camera or mobile phone had GPS, it would get confiscated. Moreover, while tourists may bring in books for personal reading, they may not bring publications that are religious or political in nature. The whole process took close to two hours before the train finally departed for Pyongyang.

From Sinuiju to Pyongyang, the scenes were mostly rice and corn fields, as well as small villages and towns. It was nearing harvest time, so there was a beautiful glow as the rays of the sun hit the green and golden stalks of rice. The rural views were immaculate, like posters from the Cultural Revolution.

The train was traveling at a slow speed. And for some reason, we made a really long stop at one of the train stations along the way. We should have arrived in Pyongyang at 5:45 P.M., but it was nearly 7:30 P.M. when we finally exited the train to set foot on North Korean soil. Our local guides were eagerly waiting for us.

We went out of the Pyongyang Train Station amidst revolutionary music. The pictures of the two former leaders Kim Il-sung and Kim Jong-il were prominently emblazoned in a place of honor above the main entrance of the train station.

A large LED screen was showing clips from cultural performances. Indeed, we were in North Korea…" — **Ivan**

* * *

"While we were backpacking in Nepal, we stayed with a local family of brothers running their small bed-and-breakfast business. They made us feel welcome as they showed us their country. Because we looked like locals, we even got to see things that only locals were privy to, such as the funeral of village elders where their bodies were being burned within a local river." — **Rambi**

* * *

"One experience that stands out was an overnight desert Bedouin trek in the Moroccan Sahara. This was during my early traveling days, when I was still a student and had more biases and insecurities due to lack of experiences. At call time, I saw that I was the only person on the tour — with a Berber guide and one dromedary (Arabian camel). My initial reaction was one of suspicion, as this was 2003, and I was apprehensive that if going out alone for this overnight trek, and camping in the desert under the stars, was safe. I decided to take the risk, and I was rewarded with an incredibly memorable experience — trekking up the giant Chebbi sand dunes, stopping at oases, drinking the most satisfying mint tea at a Bedouin camp, and helping the guide make the most mouthwatering vegetable *tagine* stew, slow-cooked in an earthen clay pot buried in the hot sand. This, along with the rest of my six-week solo backpacking journey in Northern Africa (Morocco, Tunisia and Egypt) made me discover my love for exploration and showed me that I loved and thrived when I was traveling on my own, even in places where the language, the culture and the people were so different. This was the trip that activated my lifelong thirst for adventure." — **Brian**

*   *   *

"I first learned of stilt dancers when I saw photos posted by an overlanding travel company. When my friend and I decided to go to Côte d'Ivoire, I knew I had to see a stilt dancer with my own eyes. Think about it: a man dancing on stilts! According to Lonely Planet, the stilt dancers undergo three-to-five years of intensive training before they can dance in public. 'They tell no one,' it said, 'not even their wives, what they are doing. Once initiated, they become empowered with the spirits who, during the dancing, direct their elaborate stunts.'

One morning, my friend and I traveled to the village of Silcoro, about an hour outside of Mann. We became acquainted with the villagers and in the afternoon, the ceremony began. We had no idea the people we had just met could dance like that! The stilt dancer was the last performer of the ceremony, and his moves were electrifying! When he was finished, the entire village of Silcoro cheered loudly. I felt like I was at a basketball game or a raucous pep rally! The energy was breathtaking. It was intoxicating. It was the first time I felt like I was in an episode of a travel television show." — **April**

*   *   *

"I went to Nepal in the year 2010 and I've met some locals in a party while in Kathmandu. One of them was a student named

Kris who was very active in the local community who, in turn, invited me to come visit him at his grandparents' place that weekend. The rural mountainous village located just a few hours outside Kathmandu turned out to be very beautiful. Kris and his cousins toured me around the village and introduced me to everyone who I think were also very excited to meet me. His grandparents prepared a simple feast. I loved every minute of that trip. While I was at the house, I observed that there wasn't a single photo on the walls so I gathered the family for a family portrait. I printed them when I came back to Kathmandu and had Kris send it over to his family. It was at the infancy of my traveling career and it deeply struck a chord in my heart. From then on, I tried to spend more time in the remote areas or at off the beaten track than the usual touristy sites." — **Jon**

*   *   *

"Touring the Golden Triangle in India with my aunt and her friends, we had to buy local clothes to dress up for the Taj Mahal tour and pictorial. It made for colorful and beautiful pictures!

On the way up to Amber Fort to visit the Amber Palace in Jaipur, we took the elephant ride — the friendly pachyderms lined up as one caravan when trudging up the hill towards the palace. I was supposed to ride the last elephant for the wonderful views from the tail end, but my aunt and her friend (who were on the elephant ahead of mine) asked to switch rides with me. I obliged, somewhat begrudgingly. Midway to the top, some commotion ensued behind me.

As it turned out, my elephant, a male, was "in heat" and made overtures with the female elephant behind (the one with my aunt and her friend), and sprayed urine all over them (elephant and guests included). My guide explained to me what was happening and we had a big chuckle all the way to the top — especially because I managed to stay dry and didn't get any of the pheromone-laden 'special' spray."
— **Raoul**

* * *

"In 2008, from Bamako, Mali, I flew with a friend for a one-day trip to Timbuktu, a UNESCO Heritage Site, and returned the next day (because the flights were only once-a-week). It was surprising to see this city; there was not one street with pavement and all the streets had sand. People walked and cars drove on the sand. It was sunny all over.

My friend and I visited the Djinquereber mosque, which was constructed entirely with mud, as well the prestigious Koranic Sankore University. Then we walked around sightseeing for the rest of the day, with a guide from a native tribe, to see monuments and a man-made lake at night with children playing beneath the shining moon.

We expected an early morning pick-up to return to Bamako to attend a conference. We waited and waited for our 5:00 AM ride to the airport, but the guide did not arrive. A man at the hotel offered to take us to the airport on his motorcycle, but we did not know how to take our luggage as we could not carry it on the top of our heads! We were getting nervous about

> *"We are the Pilgrims, master; we shall go. Always a little further; it may be. Beyond that last blue mountain barred with snow. Across that angry or that glimmering sea, White on a throne or guarded in a cave. There lies a prophet who can understand Why men were born: but surely we are brave, Who take the Golden Road to Samarkand." – James Elroy Flecker*

missing the plane, as the flight was just once a week and had a conference to attend in Mali. My friend started to knock door-to-door for someone who had a car. She was told that the manager of a hotel was sleeping on the roof, one floor up, and she climbed up there. The man came out of his tent naked. He agreed to take both of us to the airport once he got dressed. It was a memorable experience to see Timbuktu." — **Luisa**

* * *

"In 2002, I was having my morning excursion on a hillside in Ajjacio, Corsica (Napoleon Bonaparte's birthplace). I thought I was doing pretty good climbing uphill when suddenly, an older woman most likely in her 70s, passed by me and waved and said something in French like 'have a beautiful day' or perhaps it was 'good luck walking up'. I acknowledged with a smile (even if I did not understand her) and lifted my head towards her. It was the most pleasant exchange of laughter that I still felt in my heart. She unveiled her head cover and we had a good snapshot.

Another memorable local experience I had was in the historic site of Registan in Samarkand, and I usually request from the locals if I can take pictures of them. Well, these mothers dragged me to have myself in their pictures instead. I had a similar experience in smaller cities in Turkey." — **Henna**

* * *

"My memorable local experiences included listening to the stories of the joyful Himba ladies dancing topless in front of us as well riding a camel along the beach in Lamu island, Kenya." — **Vhang**

*   *   *

"After a day of sightseeing in Petra, my Bedouin guide invited me over for dinner with his family, and I happily obliged. We headed to his village, quite a departure from cave dwellings Bedouins used to live. I met his family, and I knew I was in for a treat when they served Mansaf, a traditional Jordanian dish. We all shared it and ate with our hands. Afterward, the host offered me a bowl of water with which to clean my hands. This unforgettable experience was not part of the tour; it was simply a gesture of their hospitality which made my trip a truly authentic local experience." — **Dondon**

*   *   *

"I had many memorable local experiences in my travels, but most of them involved hanging out with specific tribes in Africa, such as the Batwas of Burundi, the Mundaris of South Sudan, the Hadzabes of Tanzania. Journeying through Africa was an all-out assault on the senses: seeing the glorious, saffron-hued dunes of the Sahara and Namib deserts, the vibrant and colorful garments of the West African people and Muslim women, and the colorful markets across the continent; listening to the chilling roar of hungry lions in Zimbabwe; savoring the sharpness and spiciness of Ethiopian cuisine;

appreciating the ever-present dust and dirt on my skin amid the sweltering heat of the sun; and smelling the pervasive scent of the earth throughout the journey. And when licked on the face by a giraffe, the most peculiar of sensations were felt that I cannot even begin to describe. My senses were heightened. All my senses were on fire.

I was in awe of the physical strength exhibited by the local ladies. African girls and women carried heavy loads of water, firewood, merchandise and almost anything else on their heads with ease. The girls can carry loads up to 100% of their body weight. Mothers were seen simultaneously carrying babies on their backs. It's also the women who perform 60% to 80% of work in the farm. These girls and women have inspired me — not necessarily because I wanted to carry heavy loads on my head or engage in farming the way they do — but, in a way that they've demonstrated their strength and resilience. They would forever serve as inspirations to me.

I was flattered with the attention I constantly received from the locals. The African women loved my Asian hair — long, straight and shiny, and told me how beautiful I am. I had equal amount of attention from the men who were not shy to show their admiration and affection. To some, I looked like a mixed race, and that was desirable. I even had a marriage proposal from a Sudanese man. Young men from Zimbabwe, Nigeria and Togo wanted to date me. Other than being already married, the problem was that I am usually far older than they thought. I'd probably have to thank my Asian genes for that. I admit that I loved the attention, but I restricted my relations to no more than just friendships and enjoyed the chance to talk

> *"Nobody can discover the world for somebody else. Only when we discover it for ourselves does it become common ground and a common bond and we cease to be alone."* – Wendell Berry

and get to know them and, perhaps, being a little bit more playful and frisky.

My heart swelled from the goodness and kindness of the locals. I felt an indescribable warmth and joy seeing the friendly waves from the natives and smiles on the children's faces and hearing them say 'Welcome.' I was overwhelmed by the kindness and hospitality of the African people: of the taxi driver in Cairo who accompanied me on foot through the busy streets of old Cairo and looked after me as if I was his own personal guest; of the Ethiopian tour guide who offered me his hoodie jacket when he sensed I was cold; of the driver and passengers of a public van who patiently waited for me to get my visitor's visa at the border between Swaziland and Mozambique; of the security guard in our campsite who made me tea the traditional Mauritanian way when I had trouble sleeping; and, of the student in Togo who gave me his own precious repellent spray when he saw I was being bitten by mosquitoes and subsequently cooked dinner for me when I told him I wanted to try some Togolese dishes.

There was so much kindness from a multitude of people that I met during my journey. I was made to feel welcome and special. My trust in the goodness of humanity got restored experiencing the kindness of these people.

I felt the most incredible sense of freedom. Daily life was completely different in Africa. It was normal not to shower for several days and not have access to toilet facilities. Sometimes, I could only take a bath when there was a body of water available

— a stream, a river, a lake or an ocean and I could only take a shower when where there was a natural waterfall or a well. I eventually learned to do bush poos and pees like a pro. I got downright dirty without squirming and slept with sweat and dirt on my skin. I'd wash my own clothes by hand. I'd fetch water and firewood. I'd light a fire for cooking and ate what was only available. I learned to wash dishes without running water. I slept out in my tent no matter what the temperature and weather was — no air-conditioning units, no heaters nor electric fans. I learned to sleep with several layers of clothing when it was freezing. I lived without make up, hair conditioner and a proper hairbrush. I did not enjoy the discomfort of bush-camping, but for some strange reason, the experience was absolutely liberating. It made me feel alive. The basic and simple living I was forced into provided that sense of freedom." — **Riza**

What is a place that has deeply moved you and why?

"The place that moved me or rather consumed me emotionally was a situation in Gabon. We hired a driver who spoke English (French is the national language of Gabon) at the hotel who took us to the border so we can cross to Equatorial Guinea. While waiting for the boat to get us there, we saw a few men carrying a load, wrapped in blankets. We asked our driver on what's happening and he informed us that there's someone sick, a patient who was discharged from the hospital. Unbeknownst to us, the sick was loaded in the same boat where we were in. In the middle of the journey, we heard moans and cries and found out the patient took his/her last breath. Among all my trips, this experience left me with a heavy heart." — **Henna**

* * *

"As I was walking towards the New Delhi train station on my first night in the city, I was confronted by what I saw: a poor child dragging a cart that carried his sick mother. It was poverty staring at me in the face. So I took a snapshot of the moment as a powerful and humbling reminder of the human condition, which has invoked feelings of helplessness and sadness during my trip to India but has also instilled a sense of hope to make this world a better place to live.

Another destination that deeply moved me is when I visited what used to be the location of the Auschwitz concentration camp in Poland. I got this spine-chilling sensation when I stepped into a room that used to be a gas chamber. There was a deafening silence in the room as I close my eyes and try to imagine what has happened here just a few decades ago - the screams of thousands and thousands of people who were gassed during World War 2 at this site. I can only offer a prayer as I stepped out of the room. Within the area, I also saw a glass room full of shoes and another glass room filled with human hair from the people who were incarcerated in this place. It was a very horrible feeling, something I would remember for a while." — **Dondon**

* * *

"The only World War II concentration camp I have visited was Dachau, just outside Munich. Visiting the actual buildings where all the gruesome events took place was one sobering experience. I felt so moved and teary-eyed when I toured the gas chambers. I found some solace visiting the memorials, among them the Protestant Church of Reconciliation, just outside the crematoria area. I also visited the Catholic Church of the Mortal Agony of Christ, a fitting tribute and memorial since most of the prisoners in Dachau were Catholics from other countries, imprisoned for being anti-Nazi resistance fighters. After that Dachau experience, I was not sure if I would want to visit another concentration camp — only time will tell." — **Raoul**

* * *

"Petra in Jordan. I went there during my first sabbatical trip on year 2009-2010. I'm such a history buff and have always been fascinated with the old civilizations. I started my trip in Turkey, then Syria, and I definitely had to see Petra when I was in Jordan. I could still remember how it felt walking through the Siq, excitement was palpable at every turn with the hope that I would finally be able to behold the famous temple. I saw a glimpse of it, very little at the beginning and a few minutes after, it was right in front of me! I had goosebumps all over while admiring the intricately carved Treasury. The photos didn't do any justice and it was actually way larger than how I envisioned it to be. I knew of Petra from a young age through books and finally having seen it for real was a dream come true!" — **Jon**

* * *

"As a Catholic, I was deeply moved when I visited Jerusalem, Israel, as the places which would be seen thereto have already been known to me having studied in Catholic schools..." — **Jimmy**

*   *   *

"My husband and I always dreamt about visiting Libya. Under Muammar Gaddafi, it was difficult to secure a visa. When he was ousted in 2011, the country experienced a bloody civil war, and we were again on the waiting list. But then, in 2013, a door opened for a Libyan business visa and we had the chance to visit all those great Roman ruins, going even to Ghadames in the Southwest. When we left, our guide shed some tears, we probably were his last guests for quite sometime. The airport was bombed, prime minister Ali Zeidan was kidnapped, and the country sank into chaos again. We were very sad to realize all this and witness the Libyan reality post-Gaddafi!" — **Vhang**

*   *   *

"Spiritually, it would have to be the Portuguese Way Camino de Santiago, half of it taking place in Northern Portugal, with the other half in Northern Spain (Galicia). Walking 20 to 35 kilometers a day in solitude (equivalent to about 4 to 6 hours of walking) gave me more time to self-reflect than I'd ever had. Stopping from church to church, befriending many on the same route and in the same stage of life, and hearing services for *peregrinos* (pilgrims) blessed me with a more profound perspective on life. Because of these experiences, I'd like to return and do more Caminos in the future, always ending in the Cathedral of Santiago de Compostela." — **Brian**

* * *

"Kathmandu, Nepal. The place just felt like home even though we were just visiting. The people were so warm and the places were so memorable." — **Rambi**

* * *

"Potala — seems to me to be the most remote place on earth on the highest mountain in the world." — **Jazz**

* * *

"The African continent has deeply moved me in many ways. The expedition though was as much of an emotional and mental journey as it was a physical one. It had been much more than just seeing the animals or the landscapes. It had been a journey that led me to a deep self-reflection.

My soul was constantly stirred and roused. Upon reaching the summit of Dune 45 in Sossuvlei of the Namib Desert, I was spellbound by the light that had set the desert on fire as the sun rose. I wanted to freeze the moment and hold my life there. When I trekked the forsaken deserts of Sudan in search of the disintegrating Nubian pyramids, my soul was stirred. When I slept under the star-studded skies in Mauritania and witnessed the desert night sky ablaze with shooting stars, it woke up my soul and brought me to tears. The world is beautiful; life is beautiful. I am grateful that I had a chance to see this part of the world and to experience so much of it on this journey.

Finally, I had constant stimulation throughout my journey, both good and bad, some of which I sought out and some I did not. There were scary and adrenaline-pumping moments, though oddly, these were the same moments that made me feel alive inside. My experiences and emotions were rich and intense; I was roused. My energy level was high; my spirit soared... There is no other place in the world but Africa where I'd felt most alive, free and unbound." — **Riza**

* * *

"Iran, friendly people and full of history." — **Luisa**

* * *

"At the top of my list would be Iran because of the mainstream media's negative impression of it. Iran is vastly rich in ancient civilization history, people are nice, and tourist places enjoy superb peace and order. Another good choice would be most of the African countries due to their situation in general. Finally, it would be Afghanistan, my reasons being somewhat similar to Iran's." — **Andie**

* * *

"Syria. I visited this place way before the Arab Spring and the heartbreaking literal destruction of the country that followed. I was on assignment in Cyprus and me and my colleague wanted to get away for a few days, so we purchased an all-inclusive

> *"The compulsion to see what lies beyond that far ridge or that ocean – or this planet – is a defining part of human identity and success." – David Dobbs, Restless Genes*

'tourist package' that included flights, hotels and tours (this was the only way we could secure a visa) and ditched the tours once we arrived and wandered on our own. I went there with absolutely zero expectations, other than the fact that it was one of the most misunderstood countries in the world held captive by an oppressive regime. Damascus, one of the oldest inhabited cities in the world, was truly a time warp. Everywhere, I was greeted with smiles, and I felt these that these were true encounters of joy and kindness among human beings. It was haunting, talking to the people, taking their photographs. My only regret was that I wish I had stayed longer. To witness a country and people so misunderstood (and ostracized) by the rest of the world continues to be one of the greatest tragedies of our time." — **Kit**

*　　*　　*

"Absolute pristine beauty deeply moves me and I found that in the Falkland Islands, South Georgia Island, and Antarctica. Antarctica was truly magical and visiting it was a once-in-a-lifetime opportunity. Because it is so remote, it is usually the last of the seven continents travelers visit. In January 2013, as I prepared myself to step on continental Antarctica, I started thinking about this momentous occasion. I was going to hit my seventh continent! I thought about my past and my journey leading up to this exact juncture in time. I thought about when my parents arrived in the States as immigrants from the Philippines, and how they would never have imagined, even in their wildest dreams, that their first child would visit

> *"The best part about airports lies in what they symbolize. Airports are places of bookends: new beginnings and long-awaited endings, arrivals and departures, hellos and goodbyes. We start in one city to end in another hundreds or thousand miles away. You enter from a desert and exit into a blizzard. In from winter, out into summer. In from familiarity, out into something completely foreign. Or vice versa. An airport is a place of transit, and not just geographically. I wish there was some sort of time-lapse to show how people change between departures and arrivals. When I arrive back home from being away, I'm never the same person I was when I left." – Alex Brueckner*

all seven continents. I thought about how they worked tirelessly and sacrificed their own lives so that my sibling and I could receive a good education. That education essentially brought me to Antarctica, where I was pursuing world knowledge. I was, and still am, immensely grateful for this journey and my place in this world.

Antarctica was also an inspiring place, so inspiring that a few weeks later, I started hatching a bigger dream — to visit all seven continents in one calendar year. Not many people have done it, especially a Filipina. At the end of November 2013, I traveled to Rwanda, making Africa my seventh continent of the calendar year! What an incredible feeling! I don't know if this incredible goal would have been possible without Antarctica's lasting impression, but I do know that I was inspired to dream big. I realized that if you work hard, stop complaining, and take your life in your own hands, dreams do come true! You only live once and you should enjoy your precious life to the fullest!" — **April**

PART 3:

TRAVEL BUCKET LIST

C H A P T E R   7

FAVORITE PLACES

Favorite Countries

#1: Japan

60% of the most well-traveled Filipinos chose Japan as one of their favorite countries making this the top favorite destination! This is Andie and Rinell's #1 favorite country.

"My favorite is Japan for being well-versed in everything (food, transportation, ambiance, people, nature, technology, entertainment, proximity to homeland, name it all!) though expensive (major drawback)" — **Andie**

"My favorite country to visit is Japan. Nice place and culture." — **Rinell**

"Japan is in my Top 3. We love their culture and food and it's a treat every time we visit!" — **Rambi**

"Japan with its unique culture" — **Jimmy**

"Japan is, for me, one of the most intoxicating yet at the same time extremely calming places on the planet." — **Kit**

#2: Italy

Jon and Raoul's top favorite country is Italy, and it comes as the second favorite overall being chosen by 45% of the most well-traveled Filipinos.

"Italy. This is one country that has everything you can ask for – stunning landscapes from snow-capped mountains and crystalline colored lakes in the north, the dreamy verdant hills of the wine-growing regions of the central states, beautiful craggy mountains hovering over historically favorite beach destination in the Amalfi Coast, all the way to the arid region of Sicily. Italy is an amalgamation of several powerful kingdoms and city-states that are fiercely proud of their culture, traditions, and food. I'm often asked by Italians where I have eaten the best pasta in their country. It is not apparent if you breeze through just the major cities, but if you'll take it slow and assimilate with the locals, you will discover that the conservative northerners take their aperitivo seriously compared to the traditionally easygoing Sicilians. Like wine that gets better with age, Italian cities and villages have so much beauty and history with them having been founded centuries ago that anywhere you go, there's just so much to discover and learn. There are so many artistic traditions.

Having lived in Turin, I had the opportunity and time every weekend to explore new places that are beyond the usual touristy spots especially most of the northern section of Italy and as far as Florence. The rest of the southern section I visited on holidays and many more subsequent trips to the country. I haven't touched on fashion and shopping yet."
— **Jon**

"I consider major countries in continental Europe such as Italy, Spain, Germany and France as favorites. Having visited them a few times, I consider that familiarity a plus factor - along with having friends and even distant relatives who live there. Plus there are still regions and cities in those countries I have yet to explore, which motivates me to make return visits. If I have to pick one, it may have to be Italy." — **Raoul**

#3: New Zealand

40% of the most well-traveled Filipinos chose New Zealand and this is Dondon and Jazz's top favorite country.

"What I love about New Zealand is the high quality of living, people having a strong zest for life, and a lot of fascinating natural destinations and adventure sports to choose from, like in places such as Queenstown. The landscapes are among the most beautiful and scenic I've ever seen in my life - things that cinematography in movies and picture-perfect postcards are made of." — **Dondon**

"Epic landscapes, colourful cultures and laid-back charm experience. This is New Zealand." — **Auie**

#4: Spain

Spain also got selected by 40% of the group and is the most favorite country of Odette and Brian.

"Spain has 47 UNESCO World Heritage sites. I have visited all except the Altamira caves. I have been to Spain more than 52 times, explored the small cities, towns, villages. I love vibrant Barcelona.

When I seek solitude, I drive for 2 hours to Panticosa, 15 miles from the border of France. I go hiking at the Ordesa y Monte Perdido National Park in the Pyrenees mountains. Spain has a variety of 300 tapas. The best wines, delicious cuisine. Fiesta celebrations every month all year round." — **Odette**

"Spain – I first went there as an exchange student in high school, then spent one semester there during college. Because of these early formative memories, I have a fondness for and loyalty to it. Not to mention the out-of-this-world food, almost-Filipino culture, fascinating history, fun way of life (La Vida Loca!), and tremendous combination of manmade and natural sights. More recently, I trekked a segment of the Camino de Santiago, which added a spiritual appreciation for the country." — **Brian**

"Spain due to its history of being conquered by the Moors, birthplace of Picasso and other known artists" — **Jimmy**

#5: France

35% of the most well-traveled Filipinos chose France as one of their favorite countries. This is Auie's favorite.

"I think France smells not just sweet but melancholy and curious, sometimes sad but always enticing and seductive. She's a country for all senses, for artists and writers and musicians and dreamers, for fantasies, for long walks and wine and lovers and, yes, for mysteries. France, you will always have my heart." — **Auie**

> *"I can't think of anything that excites a greater sense of childlike wonder than to be in a country where you are ignorant of almost everything." – Bill Bryson*

"France due to its history and culture where one would see the bullfight arena also, palaces like the Carcassonne, Lourdes as the site of miraculous happenings through the Virgin Mary" — **Jimmy**

"The reason why we keep going to France and Mexico is the art culture." — **Henna**

#6: Australia

Australia was selected by 35% of the most well-traveled Filipinos as well.

"Our trip to Australia was one of the best experiences we have had in our entire traveling lives. It was our first trip to Australia as a couple and our, *'drumroll please,'* last continent. We officially finished all 7 continents, which is a pretty big milestone for both of us! If you are someone who loves to be on the road, then go to Australia and join a road trip adventure. It is the ultimate road trip paradise." — **Kach**

"It's magical! I love the adventure, the outback. I have many great memories there." — **Luisa**

"What is it like to travel in a country continent? Australia has so much to offer, and among my favorite picks are Sydney Harbour Bridge on New Year's Eve, Confest festival, Mardi Gras, Australian Open, Great Barrier Reef, Uluru, Kakadu National Park, Great Ocean Road, The Whitsundays, Barossa Valley, Blue Mountains, Cradle Mountain, Canberra ANZAC Memorial, St. Kilda, Bondi Beach, Surfer's Paradise, Byron Bay, Port Stephens, Sunshine Coast, Rottnest island, Kangaroo Island, Fraser Island, Margaret River, Esperance, Cape Le Grande National Park, Museum of Old and New Art, Melbourne laneways and hidden bars, and of course, the iconic Sydney Opera House. I still have a number of bucket list in the Land Down Under, and among the top in my list are Lord Howe island, The Kimberley, Broome, Magnetic island, Wineglass Bay, Burning Seed, the MCG AFL Grand Finals, and the Ghan. Beyond the places and events, there's also the people, the accent, the food, the culture, the kangaroo and koala, the laidback lifestyle, and the quality of life that makes some of the Aussie cities among the most livable in the world. To be honest, I fell in love with this country, and that's why I now call Australia home." — **Dondon**

#7: Norway

30% of the group chose Norway as one of their favorites, and this is Badong's top choice.

> *"To travel is to discover that everyone is wrong about other countries." – Aldous Huxley*

"My favorite country is Norway as I love the scenic view of the mountains and the outdoors." — **Badong**

"I love Norway for its wide-open spaces, picturesque views of the water and the sky juxtaposed to each other, for the ease and safety of travel, efficiency of the system, the cleanliness of the place. It feels I am always in communion with nature." — **Jazz**

"Norway is one of the most beautiful countries in the world. It could bless you with its charm all year round. Depending on what season you'll visit, you'll find it nothing less than extraordinary." — **Kach**

#8: Brazil

Brazil is April and Jimmy's most favorite country, and this was also chosen by 30% of the most well-traveled Filipinos.

"This is a very popular question asked amongst travelers and it is a very difficult one, especially if you have traveled to numerous countries. Each and every trip is special in its own way, but I do have my favorites. Brazil is one of them because it is very special to me. It is the first country that I traveled to as a solo traveler. Each and every time I arrive in Rio de Janeiro's international airport (GIG), I feel alive. I immediately want to dance; I want to sing! There is a certain energy in Brazil that I have not found in any other country. I've been to Brazil four times, which is an indicator of how much I love it. Because I would like to visit every country, I don't usually return to a country. Brazil offers so much diversity in its people and nature that I never get tired of it and want to continue exploring different regions each time I return." — **April**

"I would say Brazil where the image of Christ with stretched hands on the Corcovado Mountains is situated where I exclaimed during my first arrival thereto 'this could be the best sight I have ever seen'." — **Jimmy**

#9: Argentina

Argentina was selected by 30% of the group as a favorite.

"Argentina is one of my favorite countries in the world because you can trek Patagonia area with no fees to pay and has easy access to transportation. In addition, the cost of tourist living is somehow cheap, and the nature is picturesque." — **Andie**

"Argentina for its steak and gauchos" — **Jimmy**

"Argentina is one of my favorites. Firstly, it was there where I got to touch a lion. My favorite food is the beef churrasco. The culture is similar to the Philippines in some ways. I was amazed by watching locals dance the tango. I also got to visit a small part of Patagonia, when I did a number of short hikes in

Ushuaia. And of course, who wouldn't recognize the song 'Don't Cry for Me Argentina!'" — **Badong**

#10: Turkey

Turkey was chosen by 30% of the group as well.

"Visitors to Turkey are usually drawn to the mysterious East meets West allure of Istanbul, the magical landscapes of Cappadocia with its balloon spotted blue skies and the clear, luxuriously warm waters of the Mediterranean and Aegean Seas…I've visited Turkey like 8 or 9 trips already, and definitely one of my fave countries to visit in the world." — **Kach**

"I love Turkey for its archeological sites" — **Jazz**

*   *   *

20% - 25% of the most well-traveled Filipinos picked these countries as one of their favorites:

#11: Switzerland

"My favourite country is Switzerland. Cleanliness, punctuality, safety, the natural beauty and also the chocolates." — **Vhang**

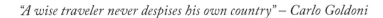

"A wise traveler never despises his own country" – Carlo Goldoni

#12: Thailand

"I can live in Thailand for their food and its Buddhist culture."
— **Henna**

#13: USA

"USA – safe, clean, organized, natural beauty, ease of travel." — **Jazz**

#14: Philippines

"I will always consider the Philippines my favorite because "It's more fun in the Philippines!". The country is best known for white sand beaches, coral reefs teeming with biodiversity, the rice terraces, colorful festivals, charming colonial towns, great food, and fun people who always smile at visitors. Plus almost everyone speaks English! It's my absolute favorite because I call it home." — **Ivan**

#15: Mexico

"Being one of the countries with the most UNESCO World Heritage Sites, Mexico has so much to offer. I guess it's the familiarity since I've been to all their states or regions and I've taken people for retreats. The cities of San Miguel deAllende, Oaxaca, Puebla (to mention a few) can rival other European towns in its charm." — **Henna**

> *"One knows that the first joy can never be recovered,*
> *and the wise traveler learns not to repeat successes but*
> *tries new places all the time." – Paul Fussel*

The most well-traveled Filipinos chose their most favorite countries and cited the basis for their selection. Based on the majority, these were the key criteria used in the order of importance:

- Natural beauty of the country including landscapes and sceneries
- Friendliness of the people
- Richness in local arts, culture, and history
- Quality of life and livability
- Great cuisine
- Safety
- Transportation efficiency and accessibility
- Diversity & broad offerings of things to do and see

There were other minor criteria cited such as affordability, weather, abundance of beaches and outdoor sceneries, environment and wildlife protection, modern cities, non-touristic destinations, and childhood associations.

* * *

"When traveling, I focus my attention on finding natural beauty, learn about local arts, immerse in the culture of the place I'm visiting, study their history, interacting with friendly local people, practicing affordable tourism, and assuring safety." — **Luisa**

* * *

> *"What you see and hear depends a good deal on where you are standing;*
> *it also depends on the sort of person you are." – C.S. Lewis*

"Natural beauty, diversity and broad offerings across the country on places of interest, friendliness of the people, safety and livability, quality of life." — **Jimmy**

* * *

"My criteria for selecting my favorite countries include possessing manmade and natural wonders; gastronomic cultures; the friendliness and liveliness of locals; childhood associations; diversity and vastness of beautiful and unique things to see; and being explorer friendly." — **Brian**

* * *

"My criteria for choosing my favorite country are the cheap cost of local living, well-preserved nature parks, ease of transportation, affordable public medical facilities, friendly locals, and low pollution levels." — **Andie**

* * *

"My favorite countries are those that mix fantasy with reality, sometimes in equal measure, both competing for my attention." — **Kit**

Favorite Beaches

#1: El Nido, Philippines (45%)

(Check out Seven Commandos Beach, Hidden Beach, Nacpan Beach, Secret Lagoon)

#2: Boracay, Philippines (40%)

"The trip that got me in high gear was an island-hopping journey around the Visayas in the Philippines which has some of the best beaches and diving sites in the world.

Great beaches in Central Philippines can be found on the islands of Boracay, Panglao, Cebu, Bantayan and Siquijor, among others."
— **Ivan**

#3: Atolls of the Maldives (35%)

(Top picks include Fihalhohi, Maafushi, Fesdu Island, Rangali Island, Full Moon Island, Lankanfinolhu)

#4: Coron, Philippines (25%)

(Don't miss Malcapuya island!)

#5: Saona island, Dominican Republic (20%)

"This is in my top 5 best beaches in the world. You know I have been telling everyone about this island/beach but no one can relate cause it's not really that accessible." — **Jon**

#6: Eagle & Baby Beach, Aruba (20%)

"Baby Beach in Aruba. We were traveling with our then two-year-old daughter Isabela and we discovered this beach where the water was only up to our ankles. We let our daughter explore and wander around and we relaxed for hours!" — **Rambi**

#7: San Blas islands, Panama (15%)

"I got engaged in San Blas islands!" — **Kach**

#8: Bora-bora, French Polynesia (15%)

"Bora-Bora comes to mind. The perfect sunset reflecting in the water with the French Polynesian mountains as your background. Wow." — **Rambi**

#9: La Digue, Seychelles (15%)

"The most beautiful from my perspective is La Digue in Seychelles although I could not freely swim thereto due to the strong waves of the azure blue waters" — **Jimmy**

#10: Zanzibar, Tanzania (15%)

"Sparkling blue waters lined with palm trees and white powdery sands, make the beaches of Zanzibar one of the most idyllic destinations of East Africa. Take long private sunset walks together at the non-tidal beaches of Nungwi and Kendwa." — **Kach**

Check out these other highly recommended destinations with amazing beaches:

- **Australia**: Whitehaven beach, Whitsundays. Hamilton Island. Esperance. Cape Le Grand National Park. Lord Howe Island
- **Brazil**: Copacobana & Ipanema beach, Rio de Janeiro. Fernando de Noronha
- **Cuba**: Cayo Jutias. Playa Varadero
- **Greece**: Elafonissi Beach, Crete. La Grotta, Corfu. Red Beach, Santorini
- **French Polynesia**: Rangiroa and Mo'orea
- **Mexico**: Playa Ruinas, Tulum. Cabo San Lucas. Cancun. Playa del Carmen
- **New Caledonia**: Isle of Pines. Luecila beach, Lifou Island. Amedee island
- **New Zealand**: Rarotonga and Aitutaki, Cook Islands
- **Panama**: Cayo Zapatilla beach, Bocas del Toro
- **Philippines**: Paradise Beach, Bantayan island. Islas de Gigantes. Balabac. Caramoan. Long Beach, San Vicente
- **Portugal**: Praia de Dona Ana, Lagos, Algarve
- **Spain**: Playa La Concha, San Sebastian. Ses Illetes beach, Formentera. Playa de Rodas, Isla de Cies, Galicia
- **Thailand**: Pa Tong Beach, Phuket. Koh Nangyuan. Ko Phi Phi
- **Yemen**: Qalansiya, Socotra

Favorite Philippine Destinations

#1: El Nido, Palawan (75%)

(you might want to consider a side trip at nearby Sibaltan)

#2: Coron, Palawan (55%)

(don't miss the Barracuda Lake)

#3: Vigan (55%)

#4: Boracay (50%)

#5: Sagada (50%)

#6: Banaue – Batad (45%)

#7: Bohol (45%)

(check out Panglao, Anda & Chocolate Hills Bike Zip Line)

#8: Siargao (40%)

(top picks include Sugba Lagoon and Sohoton Cove)

#9: Batanes (40%)

#10: Cebu (40%)

(feel free to explore Badian, Dalaguete, Moalboal, and Oslob)

> *"The whole object of travel is not to set foot on foreign land; it is at last to set foot on one's own country as a foreign land." – Gilbert K. Chesterton*

20% to 30% of the most well-traveled Filipinos picked these destinations:

#11: Port Barton, Palawan

#12: Puerto Princesa, Palawan

#13: Amanpulo, Palawan

#14: Long Beach, San Vicente, Palawan

#15: Tubbataha Reefs, Palawan

#16: Balabac, Palawan

#17: Kalanggaman Island, Leyte

#18: Siquijor

#19: Legazpi

#20: Davao

Check out some of these other highly recommended Philippine destinations:

- **Cebu, Negros & Panay**: Apo Island. Manjuyod. Danjugan. Lakawon. Bantayan island. Malapascua. Seco. Sipalay. Islas de Gigantes. Dumaguete. Sumilon. Guimaras.

- **Luzon**: Kaparkan Falls. Polillo Island. Balesin. Las Casas Pilipinas de Acuzar. Mt. Pinatubo. Kiangan Rice Terraces. Tagaytay. Babuyan. Calayan Island.

- **Mindanao**: Lake Sebu. Malamawi. Tawi-Tawi. Asik-Asik Falls. Mati

- **Mindoro, Romblon & Palawan**: Sibuyan Island. Tablas island. Apo Reef. Sablayan Zipline. Cresta de Gallo. Carabao island. Linapacan island.

- **Quezon & Bicol**: Calaguas. Caramoan. Balesin. Subic beach in Matnog.

- **Samar, Leyte & Biliran**: Lulugayan Falls. Cuatro Islas. Canigao. Sambawan

TRIVIA: In 2019, the Banwa Private Island in Palawan, Philippines became the most expensive beach resort in the world. It costs a staggering $100,000 a night!

Favorite Cities in the World

#1: Buenos Aires (35%)

#2: Barcelona (30%)

#3: Rio de Janeiro (25%)

#4: New York City (25%)

#5: Tokyo (20%)

#6: Santorini (20%)

#7: Paris (20%)

#8: Budapest (20%)

#9: Capetown (20%)

#10: Bangkok (20%)

The following cities were selected among the favorites by 15% of the group:

Berlin	**Medellin**
El Calafate	**Melbourne**
Istanbul	**Rome**
Porto	**T'bilisi**
Osaka	**Vienna**

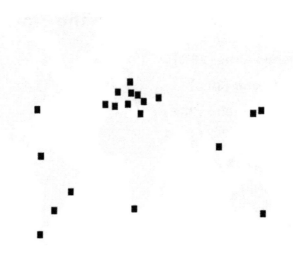

The most well-traveled Filipinos chose their most favorite cities and cited the basis for their selection. Based on the majority, these were the key criteria used in the order of importance:

- Richness in local arts, culture, and history
- Great cuisine and lots of food options
- Scenic views and natural environment
- Quality of life and livability with an easy going and relaxed vibe
- Friendliness of the people
- Safety
- Transportation efficiency and accessibility
- Cost of living
- Diversity & broad offerings of things to do and see
- Vicinity to other places of interests

There were other minor criteria cited such as being modern and massive bustling areas, great shopping, and hub for world travelers.

* * *

"Overall vibe of the city. From the people, natural environment, history, and culture. I also took into account livability, transport and food options!" — **Rambi**

* * *

"My criteria is the view, sightings and history, culture. #1 is always the view. I want to see the view. I'm fascinated with people and the culture, buildings, and stories behind it - stories of kings & queens. I'm fascinated with royalty as there's something magical about it." — **Auie**

* * *

"I love massive, bustling cities where I can flaneur all day and be completely fine with no set plans or agenda" — **Kit**

* * *

"Some are places I've lived or worked in and have affinity to because I had such a great time there; some have beautiful dramatic natural settings that I could never get tired of; all are interesting because of the culture, the people, the food, the drinks and the lifestyle." — **Brian**

"I prefer modern, vibrant, clean, rich cities that are cultural and gastronomic hotspots. Ideally places that have easy access to outdoor activities especially mountains. Fortunate to have the chance to live at some of them." — **Jon**

TRIVIA:

Do you know the New 7 Wonders Cities?

- Beirut, Lebanon
- Doha, Qatar
- Durban, South Africa
- Havana, Cuba
- Kuala Lumpur, Malaysia
- La Paz, Bolivia
- Vigan, Philippines

The intent of the New 7 Wonders Foundation is to showcase the cities that best represent the achievements and aspirations of our global urban civilization.

Top WOW Places

#1: Antarctica

#2: Mt. Everest

#3: Aurora Borealis at Kirkjufell, Iceland

#4: Pyramids of Giza

#5: Norwegian Fjords

#6: Iguazu Falls

#7: Grand Canyon

#8: Milford Sound, New Zealand

#9: Harbor of Rio de Janeiro

#10: Victoria Falls

Top Artsy Places

#1: Paris, France

"At the Louvre museum in Paris, I was eager to see the painting of Mona Lisa. I discovered more on that museum a richness of history and not just in popular art." — **Badong**

"In Paris, I prefer the smaller museums such as the Musee d'Orsay." — **Raoul**

#2: Barcelona, Spain

"I find Gaudi's art in Barcelona exceptional" — **Jimmy**

#3: Berlin, Germany

"Barcelona and Berlin are artsy cities. However the most beautiful artsy park that deeply moves me is the Vigeland Park in Oslo." — **Riza**

#4: New York, USA

"Ask most people and they tend to have the same key attractions on their list of things to do in New York; the Empire State Building, the Statue of Liberty and Ellis Island, the 9/11 memorial and at least one of the many museums, such as the Guggenheim." — **Kach**

#5: St. Petersburg, Russia

#6: Cairo, Egypt

#7: London, UK

"British Museum, Hermitage, Cairo Museum" — **Jazz**

#8: Seoul, South Korea

"Seoul is a creative hub and it manifest on their building architectures, interiors, quaint cafes, street arts etc. The Korean fashion in particular has caught the world by storm. I also like how they were able to embrace modernity without sacrificing their heritage." — **Jon**

#9: Athens, Greece

#10: Cinque Terre, Italy

> *"Beautiful things don't ask for attention."* – James Thurber,
> *from the film The Secret Life of Walter Mitty*

Most Picturesque Places

#1: Swiss Alps, Switzerland

"I like the almost perfect landscape of Lauterbrunnen region - snow-capped mountains over a lush green terrain, interspersed with cute cabin lodges sitting on the slope. And down in the valley floor, you can find more houses, restaurants and quaint shops, a snaking river, and the famous waterfalls on the valley walls." — **Jon**

"I saw my picture perfect scenery at the Glacier Express in Switzerland during the scenic ride though the alps from Zermatt to St. Moritz" — **Luisa**

#2: Geiranger and Lofoten, Norway

#3: Himalayas, Nepal

"I went to see Everest in May 2013, and it was a good month to visit climate-wise. For those of you who don't have much time to do the full Everest Base Camp, which will require over ten days, you can opt for a five-day hike roundtrip to the town of Namche Bazaar from Lukla airport, and you will be able to see Mount Everest. While I was there, I took this amazing 360 panoramic shot of the Himalayas from left to right in one photo: Tabuche, Nupche, Everest, Lhotse, Peak 30th, Ama Dablam, Thamserku, Kusum Kanguru, Konde, Thame, Khumbila. It was magnificent!" — **Dondon**

#4: Atacama desert, Chile

"The Atacama Desert is said to be the driest place on Earth. In some areas, the otherworldly appearance resembled that of Mars which makes the desert a favorite filming location for Mars scenes. It was surprising to see towns and cities built in the middle of the desert." — **Ivan**

> *"Though we travel the world over to find the beautiful, we must carry it with us, or we find it not."* —Ralph Waldo Emerson

#5: Canadian Rockies, Canada

"I'm a photographer but I find that landscape photography doesn't do anything to me emotionally. But if I was to give an answer, I would say the Canadian Rockies." — **Kit**

#6: Isle of Skye and Scottish Highlands, UK

#7: Amalfi Coast, Italy

"No trip to the Amalfi Coast is complete without hiking the appropriately named Sentiero Degli Dei (Path of the Gods), at least for me. Tiberius, the second Roman emperor and one of its greatest general frequented the place while living in nearby Capri. The hike starts in the village of Aregalo, passing through cliffs, vineyards, uniformly white-painted Italian villages, the glorious Amalfi coastline, and culminating in stunning Positano. Very beautiful hike and actually really easy. Just plenty of stairs towards the end." — **Jon**

#8: Cappadocia, Turkey

"Two years later, on the exact Saturday she left us, Mom appeared as a rainbow in the sky amidst 73 hot air balloons in Cappadocia. What an amazing way to celebrate life!" — **April**

#9: Oia, Santorini, Greece

#10: Deadvlei, Namibia

"The place where I consider a picture perfect scenery is Deadvlei inside the Namib-Naukluft Park (UNESCO) near Sossusvlei in beautiful Namibia." — **Vhang**

Most Exotic Places

#1: Easter Island

#2: West & Central Africa

"Africa offers many exotic places for me such as most countries in West and Central Africa." — **Riza**

#3: Bhutan

#4: Turkmenistan

#5: Pakistan

"I have great memories of Pakistan during my 2018 trip with Meg Pipo. Though the country gets featured in international news for all the wrong reasons, we were far from being deterred from visiting the country. We were glad we took the plunge and had an experience like no other. I remember the truck art, a popular regional, traditional decoration in Pakistan, with lively colors that have inspired ornate floral designs and calligraphy. We visited Lahore, which is a big city, and the people are super friendly and hospitable. People showed us around and stuffed us with more food than we've eaten in our entire lives. We made some great friends in the process and hope to head to Lahore again soon." — **Odette**

#6: Mongolia

#7: Tibet

#8: Afghanistan

"I have this visual image of Afghans in my mind as I saw them from photos and the media. To be specific, the women are covered in hijab from head to toe. So, when I went to the Wakkan Valley in northern Afghanistan, I was shocked that that isn't the case in the part of the country. Majority of the residents were practicing the Ismaili sect of Shia Islam which is probably the most liberal in the world. The women are wearing almost western clothing and so are the men. There's not even a mosque is sight but only some modern community center which serve as their gathering place." — **Jon**

#9: Transnistria

"Transnistria, the 'country that doesn't exist', is a product of the messy breakup of the Soviet Union, when a part of the newly formed Republic of Moldova that wanted to stay Russian rebelled against Romanian domination from Kishinev (Chisinau) and decided to go its own way. Wander the streets of Tiraspol and explore the country's Stalinist charm." — **Kit**

#10: Tanna Island, Vanuatu

"Is it possible to be that close to see lava from an active volcano's crater? It's one of those questions you'll have until you reach the top of Mount Yasur in Tanna, Vanuatu. Finding out that it is the most accessible active volcano, I journeyed with a local host and trekked towards the crater filled with excitement. We reached our destination before the sun came down. It was windy and volcanic ashes were blowing towards us. It, however, did not deter me from looking down to see red, bright lava boiling, flowing, and erupting in magnificence." — **Dondon**

7 Wonders of the World 1st Alternative Group Choice	7 Wonders of Nature 1st Alternative Group Choice
#1: Moai Statues, Easter Island, Chile	#1: Himalayas, Nepal
#2: Bagan Temples, Myanmar	#2: Galapagos islands, Ecuador
#3: Angkor Wat, Siem Reap, Cambodia	#3: Zhangjiajie National Forest Park, China
#4: Abu Simbel, Luxor, Egypt	#4: Great Barrier Reef, Australia
#5: Acropolis, Athens, Greece,	#5: Perito Moreno Glacier, Argentina
#6: Burj Khalifa. Dubai, United Arab Emirates	#6: Grand Canyon, USA
#7: Ruins of Baalbek, Lebanon	#7: Fjords in Norway

7 Wonders of the World 2nd Alternative Group Choice	7 Wonders of Nature 2nd Alternative Group Choice
#1: Hampi, India	#1: Blue Lagoon, Iceland
#2: Samarkand, Uzbekistan	#2: Salar de Uyuni, Bolivia
#3: Tikal, Guatemala	#3: Dallol Crater, Danakil Depression, Ethiopia
#4: Terracotta Army in Xi'an, China	#4: Volcanic Lake at Mount Nyiragongo, Democratic Republic of Congo
#5: Nazca Lines, Peru	#5: Gates of Hell, Turkmenistan
#6: Hagia Sophia, Turkey	#6: Serengeti, Tanzania
#7: Batad Rice Terraces, Philippines	#7: Cappadocia, Turkey

TRIVIA:

Do you know the New 7 Wonders of the World?

- Great Wall of China
- Chichén Itzá in Mexico
- Petra in Jordan
- Machu Picchu in Peru
- Colosseum in Italy
- Taj Mahal in India
- Christ the Redeemer Statue in Brazil

The Great Pyramid of Giza, Egypt is the only one of the seven wonders of the ancient world that still stands today.

Do you know the New 7 Wonders of Nature?

- Amazon Rainforest and River
- Halong Bay in Vietnam
- Iguazu Falls in Argentina & Brazil
- Jeju Island in South Korea
- Komodo Island in Indonesia
- Puerto Princesa Subterranean River in Philippines
- Table Mountain in South Africa

Most Challenging Places to Go

#1: Libya

"The most challenging place I have traveled to so far is Libya. Currently, they do not issue tourist visas, only business visas. So one has to work with a fixer in order to obtain a business visa. It hadn't been straightforward for me. My application was initially rejected and my fixer had to pull strings in order to get my application reconsidered and approved. This was also expensive." — **Riza**

#2: Papua New Guinea

"It's March 2015, and this is my last week in Port Moresby, Papua New Guinea, after being based here for almost half a year. What makes this place challenging is the rampant presence of random opportunistic violence. I could not even walk around the streets of Port Moresby unaccompanied. Though it's probably one of the least safe places I've ever been around the world, it's also one of the most fascinating to explore, perhaps because some areas are so exotic, raw, and untouched. I came to adapt and learn to appreciate the place, the people, and the culture. I have been privileged to have experienced PNG, the land of the unexpected." — **Dondon**

#3: Drake Passage to Antarctica

"I believe Antarctica because of the many challenges especially if one would cross the Drake Passage which is the converging point where the Atlantic, Southern and Pacific oceans meet thus it is a place where passenger invariably would develop seasick and vomit" — **Jimmy**

#4: Democratic Republic of Congo

"Collectively, Central Africa (9 countries) is the hardest region to travel in the world - expensive visa (DRC - 324 usd, South Sudan - $ 200, Equatorial Guinea - $223), exorbitant flight tickets (average $ 200 for an hour flight), horrible land border crossings, transport, expensive accommodation etc. This region includes the relatively tough as nails to get a visa country of South Sudan, Angola, Cameroon, and Democratic Republic of Congo (Congo - Kinshasa)." — **Jon**

#5: South Sudan

#6: Angola

#7: Cameroon

#8: Equatorial Guinea

#9: Yemen

"We breathed a sigh of relief after taking off and reaching a safe elevation. There had been shootings the night before our flight, then bazooka firings the following morning on our way to the airport. An armed conflict was nearby that each bazooka explosion shook the road and airport ground. I was worried our flight would be cancelled, but Yemenia flew as scheduled to our relief. Sadly, Yemen remains to be a country in active war and with ongoing armed conflicts." — **Riza**

#10: Afghanistan

"I don't know how to visit the country since the Taliban has taken over Afghanistan in August 2021." — **Dondon**

> *"When a man is a traveller, the world is his house and the sky is his roof, where he hangs his hat is his home, and all the people are his family." – Drew Bundini Brown*

Preferred Places to Live

#1: Berlin, Germany

"I would love to live in Berlin because of the creative vibe of the city." — **Kit**

#2: San Diego, California, USA

"La Jolla, or Santa Barbara, California – mild climate year-round, small city, safety, ease of getting around, cosmopolitan city, natural beauty" — **Jazz**

#3: Melbourne, Australia

"You have to live in Melbourne for a while to truly understand why it's one of the most livable cities in the world. It grows on you." — **Dondon**

#4: Barcelona, Spain

"I have long wanted to live in Barcelona and Cape Town. I love places with fresh fish, good wine and warm weather (or beautiful sunshine). I love their proximity to the mountains and sea." — **Riza**

#5: Medellin, Colombia

"My picks change every year. My criteria would be a place that is vibrant, clean, developed with plenty of cultural activities and has good dining options. Ideally places that have easy access to outdoor activities especially mountains. Through the years my choices were the following and fortunately I had the opportunity to live at some of these places - Munich, Melbourne, NYC, Rio de Janeiro, Medellin, Bandung, Seoul, Almaty." — **Jon**

#6: Lucerne, Switzerland

"Lucerne, Switzerland because quality of life is high, good infrastructure and health care, nice scenery and lots of opportunities to do in the spare time." — **Vhang**

#7: Madrid, Spain

"I'd like to live in Madrid again. I have a few friends there, making the transition easy, love the food, the nightlife, the history, the way of life, the sights, the people and the culture. Also, ample beautiful sites for historical sightseeing, hiking, beaching and eating are just a short train ride away." — **Brian**

#8: San Francisco, USA

"I feel fortunate to live in the Bay Area in California where the weather is gorgeous almost all year round. Outside of the US, I would consider Frankfurt or Berlin, as Germany is central enough to springboard to other cities around Europe. I can do some consulting work while visiting (or re-visiting) places in Europe I have not visited before." — **Raoul**

> *"Every one of a hundred thousand cities around the world had its own special sunset and it was worth going there, just once, if only to see the sun go down." —Ryu Murakami*

#9: New York City, USA

"Paris, Berlin, New York and California" — **Henna**

#10: Manila, Philippines

"No place I'd rather be than Manila, Philippines. Access to local cuisine and to an affordable cost of living are our priorities." — **Rambi**

> **TRIVIA:** The Economist Intelligence Unit releases an annual global livability index which ranks 140 cities worldwide based on over 30 qualitative and quantitative factors across five categories: stability, healthcare, culture and environment, education and infrastructure. This becomes the basis for determining the world's most livable cities.

Top Bucket List

#1: Sub-Saharan Africa

"I've officially entered visa hell. Every African country I haven't been to yet requires a visa except for Equatorial Guinea, CAR, Madagascar, and Mauritius. The logistics of flight paths and visas for the remaining 10 African countries left for me to visit are enough to drive someone insane!" — **April**

"I'd like to do an Overland Africa trip hitting as many Sub-Saharan countries as possible." — **Brian**

#2: Antarctica, South Georgia and the Falklands

"Ever since I started dreaming of traveling the world, there have been a small number of places that lived at the top of my mental list of, 'places I have to see before I die!'. I traveled to one of the places at the very top of that list, Antarctica, which took us from the southern fjords of Chile and across the formidable Drake Passage to the Antarctic Peninsula where we swam in icy waters and hiked across volcanic islands among huge 'waddles' of penguins!" — **Kach**

"Antarctica! The last continent has been a long-time goal of mine and I am planning to visit it this 2021." — **Rambi**

#3: The Stans

"The strictest among the stans is Turkmenistan - first I had to get a letter of invitation provided by a tour company then join

their tour. Uzbekistan Tajikistan no invitation required joined tours Kyrgyzstan Kazakhstan I stayed w families solo tour Pakistan 2 persons DIY tour. All stans are a must see especially Uzbekistan"
— **Odette**

"Very high in my list is to visit the 7 Stan countries: Afghanistan, Pakistan, Kazakhstan, Kyrgyzstan, Tajikistan, Turkmenistan, and Uzbekistan." — **Dondon**

#4: Trans-Siberian

"No English-speaking Russians during my Trans-Siberian journey (even the Provodnitsa); tourists I've met can't communicate that well also." — **Andie**

"The longest period of my Trans-Siberian trip was the four-day journey from Irkutsk, Siberia, to Moscow. The train stops for around 15 minutes to drop off and pick up passengers along the route. I have somehow accustomed myself to living in the cabin with the convertible bed/seat, content with the complimentary hot water so I can eat my noodles and not having to take a shower for this long. At some point, I had to do something different and interesting to break the monotony. So during the stop-over in the city of Krasnoyarsk, I saw a couple of Russian soldiers, and I summoned up the courage to ask for a photo with them through hand signs because of the language barrier. They happily obliged!" — **Dondon**

> *"This was how it was with travel: one city gives you gifts, another robs you. One gives you the heart's affections, the other destroys your soul. Cities and countries are as alive, as feeling, as fickle and uncertain as people. Their degrees of love and devotion are as varying as with any human relation. Just as one is good, another is bad."* – Roman Payne

"Did extensive research last year start from Vladivostok Trans-Siberian railway have to pay USA $290 for visa…was scheduled to leave Sept 2020 until Oh No pandemic. I was in Irkutsk Olkhon island in Lake Baikal 2014. Hopefully world opens up soon." — **Odette**

#5: Madagascar

"I made it to the capital of Madagascar that locals also call Tana. I was surprised when I arrived because I exactly looked like the locals! During my whole stay, everyone just thought I am a local until I start talking. Well, even if Madagascar is part of the African continent, most of the people here are of Malaysian descent. They moved here a very long time ago, but because of their Malaysian descent, they look very much like Filipinos!" — **Kach**

"My Top Bucket list is Scandinavia, Madagascar, Seychelles, Ethiopia, South Georgia and the Falkland Islands." — **Vhang**

#6: Lord Howe Island

"When I entered Lord Howe I felt that the terrain and the accompanying flora and fauna are substantially diverse as the scene is likened from another planet." — **Jimmy**

'This is in my bucket list! Thanks for sharing." — **Riza**

#7: Socotra, Yemen

"Landed in Socotra after a long flight from home. With no internet for 7 days: Spending the day relaxing having lunch, beaches, fisherman catch of the day and enjoying the beautiful sunset" — **Luisa**

"I have Syria, Iraq and Yemen left to complete the Asian countries" — **Andie**

#8: The Caucasus

"I love geopolitics and one region where I learned so much on the ground rather than from readings is the Caucasus. Such a complex history, so much violence, and still obscure boundaries. By just looking at the map, you'd see 3 countries but in reality there are 3 countries and 4 territories - 6 of them you need a visa or go through different entry regulations." — **Jon**

"My bucket list includes the 5 Stans, 3 Caucasus countries, countries of Western, Eastern and Central Africa, Mongolia, Manchuria, Siberia" — **Jazz**

#9: Tibet

"Exotic places I've visited include Bhutan, Bolivia, Tibet" — **Jazz**

"Tibet and Egypt are on my top bucket list." — **Henna**

#10: Wakhan Corridor and Pamir Highway

"I embarked on a 2-month trip around Central Asia and my first stop is the road trip from Tajikistan to Kyrgyzstan and the latter being known as The Land of Celestial Mountains and is famous for the Pamir Mountain Ranges where the Pamir Highway lies."— **Kach**

"My bucket list includes Trans-Siberian Railway or the Wakhan Corridor/Pamir Highway"— **Kit**

CHAPTER 8

FAVORITE TRAVEL EXPERIENCES

Favorite World Festivals and Events

#1: Carnival of Rio de Janeiro

#2: Oktoberfest, Munich, Germany

#3: Mardi Gras, New Orleans, Louisiana, USA

#4: Philippine Visayas Festivals: Ati-Atihan, Dinagyang, Pintados and Sinulog

#5: Burning Man

#6: Day of the Dead in Mexico

#7: Full Moon Party in Koh Pha Ngan, Thailand

#8: Holi Festival, India

#9: Running with the Bulls in Pamplona

#10: Street Parade, Zurich, Switzerland

*   *   *

"I kept thinking while witnessing the spectacle of a thousand Bandjoun natives donned in colorful traditional customs stomping their feet on a sandy patch of ground creating a joyful eclectic sound. In the center, dozens of drummers earnestly creating festival melodies that even non-dancers like me can't help but gyrate at its sound. I'm the only foreigner in the stand, even that I've acquired African complexion in the few months that I'm traveling in the continent, I still stand out wherever whether I like it or not. The cameraman, aware of the presence of the solitary mzungu, would occasionally flash his camera towards our direction, hoping to catch an image of me obviously enjoying every moment of the fest. A few minutes later, a loud chorus erupted in the stand when the King joined the dancers. Everybody suddenly put up their phones in the hope of capturing the King's joyous celebration. He passed our area, smiled, and pointed at me. I wave, gave a polite bow but smiling through my ears oblivious of everyone's attention. And on he went. It's one of those moments in my traveling life that I will remember for a lifetime." — **Jon**

"Imagine living for nine days in a temporary city in a desert with over 75,000 other people; where creativity and innovation live through the art cars, costumes, and playa art; where you can wear as much or as little as you want; where the only thing money can buy is ice and coffee; where you're off the grid as there's almost no reception; where you get to design how you want to participate in life through radical rituals you choose; it's an immersive experience where you find comfort and joy as you discover your Playa name in the process. These words have significance to Burners like me: Sandstorms. Bicycles. LED lights. Temple. Mutant Vehicles. Robot Heart. Mayan Warrior. Porta-Potties. Trash Fence. Tutu Tuesday. Human Carcass Wash. Decompression. Of all my world travel experiences, Burning Man is, hands down, at the top of my list!" — **Dondon**

Highly Recommended World Festivals and Events

- ❑ Arirang Mass Games, Pyongyang, North Korea

- ❑ Harbin International Ice & Snow Sculpture Festival, China

- ❑ Pingxi Festival; Lantern Festival in Kaoishiang, Taiwan

- ❑ Chinese New Year in Beijing, China

- ❑ Day of the Sun in Pyongyang, North Korea

- ❑ Confest in New South Wales, Australia

- ❑ World Expo

- ❑ Diwali, India

- ❑ Durga Puja, India

- ❑ Goroka Sing-sing Festival, Papua New Guinea

- ❑ Hamar Wedding Ceremony, Omo Valley, Ethiopia

- ❑ Fringe Festival, Edinburgh, Scotland

- ❑ Floriade Expo, Netherlands

- ❑ Hajj, Mecca, Saudi Arabia

- ❑ Flower Festival, Madeira, Portugal

❑ Calgary
Stampede,
Calgary, Canada

❑ Naadam
Festival,
Ulaanbaatar,
Mongolia

❑ Malabo Body
Painting
Festival,
Equatorial
Guinea

❑ Midsummer,
Finland

❑ Holy week
Zaragoza, Spain

❑ Carnivale of
Venice, Italy

❑ Songkran
Festival,
Thailand

❑ Thimpu
Tshechu
Festival, Bhutan

❑ Timket and
Meskel,
Ethiopia

❑ Tomatina,
Buñol, Spain

❑ Comic-Con,
San Diego,
USA

❑ Voodoo Festival,
Benin and Togo

❑ Tournament
of Roses,
California, USA

❑ Naghol Land
Diving Festival,
Vanuatu

❑ Yi Peng and
Loy Krathong,
Chiang Mai,
Thailand

Favorite Celebration Parties

Carnival and New Year's Eve in Rio de Janeiro, Brazil

New Year's Eve in Times Square, New York, USA

New Year's Eve in Sydney Harbour, Australia

New Year's Eve in Paris, France

New Year's Eve in Dubai, UAE

Carnaval Bahia in, Bahia, Brazil

Cologne Carnival in Cologne, Germany

Mardi Gras Parade in Sydney, Australia

Mardi Gras in New Orleans, USA

St. Patrick's Day, Dublin & NYC

* * *

"It's been a tradition for our family to spend New Year's in a big city. In the past two years, we've done Milan and Seoul. This year, we wanted to experience Paris! We chose to do the countdown in Arc De Triomphe, and it did not disappoint!" — **Rambi**

> *"... then with the arrival of noisy helpers the scene became one of riotous carnival. For they carried boxes of coloured balls, bales of scarlet and yellow bunting, baskets laden with glittering tinsel, trumpets painted silver and vermilion, dolls in vivid muslin dresses, stars and medallions, tops and skipping ropes, and tumbled them in festive profusion over baskets and chairs." – Winifred Holtby*

"One of the highlights of my trip to Rio de Janeiro during the Carnival was going to the Favela Funk Party. It took a long drive to the foot of Vidigal mountain, and then we rode on motorbikes driven by locals to the top. It was a bit scary at first as it was favela territory (Brazilian shantytown), but the area has been prepped in time for the Carnival events. When we reached the top, this huge party organized by a popular hostel was already full swing. After a whole night of dancing, we were greeted by a spectacular sunrise with a view of Ipanema beach below us. What a way to end the party!" — **Dondon**

Memorable Party Destinations

#1: Brazil: Rio de Janeiro and Sao Paolo

"Brazil. They are party animals. We had this charter cruise in Sao Paulo, Brazil. The party started when people got onboard from 11 in the morning, and it finished the next day. The drinks were flowing all over the place non-stop." — **Badong**

"I like the party scene in Rio. From the posh disco houses in Ipanema and Copacabana to the dusk till dawn street party in Lama, there's options for every kind of traveler, budget wise. The Cariocas really know how to party." — **Jon**

#2: Spain: Ibiza, Barcelona, Gran Canaria, Tenerife

"I was in Ibiza, Spain, and partying hard for my upcoming birthday was on the top of my list! Partying in the largest nightclub in the world with a world-famous DJ, CHECK! An all-nighter party in a warehouse superclub, CHECK! Relaxing by the beach in the mornings to recoup energy, CHECK! Hitting a massive foam party until sunrise before an early flight the next day, CHECK! Boarding the airplane with wet shoes straight from the foam party was ONE BIG CHECK! Ahh, the life!" — **Dondon**

"Barcelona, Spain. Some Air Force buddies and I spent 4 sleepless nights partying it up by the beaches in Barceloneta. Fun times!" — **Rambi**

"I feel it would be that in Tenerife in Spain" — **Jimmy**

#3: Greece: Mykonos

#4: Germany: Berlin

#5: Hungary: Budapest

#6: Serbia: Belgrade

#7: Netherlands: Amsterdam

#8: Thailand: Bangkok, Phuket and Ko Pha Ngan

"I have to say that the New Year Countdown Full Moon Party in Haadrin Beach at Ko Phangan, Thailand, has been the wildest party I've ever been to!" — **Dondon**

#9: USA: Las Vegas, South Beach in Miami

#10: Belize: Caye Caulker

"I am not a big party person, probably the Caye Caulker in Belize was my last party place." — **Riza**

Favorite Music Festivals

Tomorrowland in Boom, Belgium

Glastonbury Festival, Sommerset, England

Coachella, Indio, California, USA

Djakarta Warehouse Project, Jakarta, Indonesia

Eurovision

Eurovision

Tomorrowland   Glastonbury

Coachella

Djakarta
Warehouse Project

Favorite World Sporting Events

#1: Soccer/Football tournaments (FIFA World Cup)

"Definitely the World Cup in Brazil. We watched live matches in Manaus, the gateway to the Brazilian Amazon. Brazil + the World Cup = a fun-filled atmosphere!" — **Brian**

#2: Tennis Grand Slam (Australian Open, US Open, Wimbledon, French Open)

"I have not watched too many sporting events outside of my home base. But I consider watching US Open tennis in New York always a fun experience, being a tennis fan." — **Raoul**

"I've been to the Australian Open 2008/2010, French Open 2011, US Open 2013/2016, and Wimbledon 2019. I've finally completed all the 4 Grandslam as a spectator. It's a huge personal achievement for me as an avid tennis fan." — **Jon**

#3: Olympic Games

TRIVIA: There are 206 National Teams that participate in the Olympics, comprised of the 193 U.N. countries plus American Samoa, Aruba, Bermuda, British Virgin Islands, Cayman Islands, Cook Islands, Guam, Hong Kong, Kosovo, Palestine, Puerto Rico, Taiwan, and U.S. Virgin Islands.

"I was visiting London during the 2012 Summer Olympics and my friend was prodding me to attend the Closing Ceremonies. While we were waiting for the online tickets to be released for the locals, I was having second thoughts because of the price. My friend, however, kept reminding me that this was an opportunity of a lifetime. Eventually, I caved in and got myself a ticket. It was amazing just being inside the stadium, even more amazing to get such good seats, close to the Olympic flame. The best part of it all was getting to meet the Australian flag bearer Malcolm Page, and as a bonus, he let me touch his Olympic gold medal. Now that's an experience truly worth every cent!" — **Dondon**

#4: NBA Games in USA

"Being a sports fanatic, my best memories would be watching the NBA Finals in Cleveland, Ohio as I watched Lebron and Steph Curry battle it out as Golden State won their third championship. Another highlight would be watching the Winter Olympics in Pyeongyang, South Korea as the US Hockey team took Gold! The most recent experience was in Lyon, France where the USA Women's Soccer Team won the World Cup!" — **Rambi**

#5: Bull fighting in Spain

#6: Formula 1 Grand Prix

#7: Nadaam festival in Mongolia

"No doubt the Nadaam festival in Mongolia, a 3-game sporting event on wrestling, horseracing and archery held during midsummer. Also, a local wresting event of the Mundari tribe in Terekeka State, South Sudan." — **Riza**

#8: Wrestling (Cholitas Wrestling in La Paz, Bolivia and WWE in USA)

#9: Tour de France

"On my second visit to Paris, it was summertime, and little did I know that the Tour de France was coming to a finale the following day. I had to decide on a spot where I can watch closely as the pelotons arrive, so I chose to be at Rue de Rivoli near the side of Champs Elysees. I thought I'd be able to get a better shot of taking photos on a much narrower road. True enough, witnessing the race unfold up close until the last lapse was quite an excitement!" — **Dondon**

#10: Human Tower Competition in Tarragona, Spain

"The highlight of my visit to Tarragona in 2019 is of course the Human Towers. To my delight, my friend Oscar Hijosa Milà messaged me, 'Hi there Don, Bring comfortable clothing this evening. You are going to join us in human castle training.' I ended up participating during the training session of his award-winning team, Colla Jove Xiquets de Tarragona!" — **Dondon**

Favorite World Music

#1: Latino music (Reggaeton, Salsa, Merengue, Samba, Tango)

"In South America, I enjoy watching couples dance the tango in public squares in Buenos Aires and Montevideo." — **Raoul**

#2: EDM, Deep House, Technotrance

"One of my most memorable EDM concerts was watching David Guetta perform during the 2013 Zurich Street Parade. I was in the front row, right in the center, directly in front of him!" — **Dondon**

#3: Afro beat

"I love to dance and have enjoyed it most when I get the chance to dance with an ethnic tribe. Some examples are dancing with the Tamberma women in Northern Togo, the Batwa people in Burundi, the Mundari people in South Sudan, the Cholistanis in Pakistan." — **Riza**

> *"The curious beauty of African music is that it uplifts even as it tells a sad tale. You may be poor, you may have only a ramshackle house, you may have lost your job, but that song gives you hope."* – Nelson Mandela

#4: Traditional Music including chanting

"I like the music culture of West Africa. I got acquainted with it when my French roommate in a hostel in Bamako told me that he's been exploring that region for years to learn the traditional instruments and incorporate it on his craft. I joined him for a beer at a place that has nightly local performances and I was hooked. I also like the cultural performances and music of Ethiopia. You could always find a restaurant that offers those kinds of shows which you could enjoy while having dinner at every major tourist cities like Addis Ababa (Yod Abyssinia) and Gondar (The Four Sisters)." — **Jon**

#5: Ibiza Chill-out music

"You know those Cafe Del Mar Buddha Bar chill out music that was ubiquitous during the 2000's? It sounds horrible in an elevator, but when it's 7 am and you're watching the sunrise over the Mediterranean after an intense night and you smell the ocean breeze…it's sublime." — **Kit**

Favorite languages

#1: Spanish

"Spanish and French, they sound so pleasant and romantic." — **Luisa**

#2: French

#3: Portuguese

"I love Spanish so I studied it. I think Brazilian Portuguese and Russian are beautiful to the ear." — **Kit**

#4: Italian

"It is understandable enough as a Romance language – having a passing level of fluency in Spanish - while intriguing enough to challenge me to parse and translate. The more I visit Rome and Italy, the more I get by with Italian – especially in places away from the big cities, where not too many people speak English." — **Raoul**

#5: Persian

"It is a language rich with history, literature and tradition. It achieved great height during the peak of the Persian empire which at some point covered a third of the world thereby leaving vast influences to the Turkish, Arab, Central Asian, and Indian culture of today. It remains the national language of Iran, Tajikistan, and Afghanistan." — **Jon**

Favorite spots to see beautiful people

#1: Medellin, Colombia

#2: Rio de Janeiro, Brazil

#3: Moscow, Russia

#4: Minsk, Belarus

#5: Vilnius, Lithuania

#6: Portofino, Italy

#7: Nis, Serbia

#8: Norway

#9: Greece

#10: Spain

Favorite Wildlife Experiences

#1: Galapagos

"Seeing wildlife up close and not in a zoo changed how I traveled. The first time I encountered wildlife I was in a region called the Pantanal in Brazil, and I saw so many beautiful birds. I do not consider myself a bird enthusiast, but the variety impressed me. I also caught my first and only piranha and saw an anteater. Since then, I have made it a point to see as much wildlife that I can. During my time in Africa, I visited more than 10 game parks and saw four of the Big Five animals. The leopard is still elusive for me! I've seen the mountain gorillas in Uganda and Rwanda, and had the most fantastic experience as one of the gorillas was very curious about my purse and kept grabbing it. I can still picture her hand touching me! I've been blessed to see sea lions and turtles endemic to the Galapagos Islands; fed a stingray in the Bahamas; swam with a wild whale shark in Djibouti; and swam with the jellyfish in Palau. I've seen wild polar bears in Churchill, Canada; and five types of penguins in the Falkland Islands, South Georgia Island, and Antarctica. I have a short video of King penguins on South Georgia Island that I frequently show to friends and acquaintances because I want them to see how incredible the sheer number of penguins there is in such a small area of that island."
— **April**

#2: Tanzania's Serengeti and Ngorongoro crater wildlife

#3: Kenya's Masai Mara wildlife

#4: South Africa's Kruger National Park wildlife

#5: Brazil's Pantanal and Amazon Rainforest wildlife

#6: Antarctica and Falkland Islands' penguins

#7: Uganda and Rwanda's gorillas

#8: Madagascar's lemurs

"I've spotted an Indri, the largest living lemur (about 3 feet), in the eastern rainforest. There used to be gorilla-size lemurs in this rainforest but they were hunted and now they are all gone." — **Riza**

#9: Australia's Outback wildlife and Great Barrier Reef's marine life

#10: Caribbean's stingrays

#11: South Africa's great white sharks

#12: Philippines' tarsiers and whale sharks

I am lucky to have had the opportunity to have close encounters with the whale sharks during my visit in Donsol, Sorsogon, as well as in Oslob in Cebu island, Philippines. They come there in groups during mornings to feed, so it wouldn't be so uncommon at all to see half a dozen whale sharks in Oslob. Some rules of engagement include not touching them and not using sunscreen. It was an awesome experience to swim beside these gentle giants and have a few underwater shots of them. One time, I jumped off the boat and almost landed right on top of this car-sized whale shark - that freaked me out for a moment." — **Dondon**

#13: Pacific Island's mantarays

#14: Palau's jellyfish

#15: Indonesia's komodo dragons

#16: Svalbard's polar bears

#17: Borneo's orangutan and proboscis monkeys

#18: China's giant pandas

#19: Tonga's humpback whales

#20: USA's Yellowstone National Park's wildlife

Unusual World Dishes We've Tried

#1: Fertilized duck egg (Balut) in Philippines

#2: Fermented shark (Hakari) in Iceland

#3: Crocodile skewer in Australia

#4: Guinea Pig (Cuy) in Cusco, Peru

#5: Horsemeat in Central Asia

#6: Worm in Africa

#7: Tarantula in Cambodia

#8: Dog soup in North Korea

#9: Rat and frog in Vietnam

"I joined a Reality TV show in Vietnam (and won), and the show dared me to eat a rat and a frog, among others. NEVER again." — **Kach**

#10: Jumping shrimps in Phayao Lake, Thailand

#11: Frog legs in Nuuk, Greenland

#12: Fermented horse milk in Kazakhstan

"Ah, Kazakhstan. The horsemeat and the intestines were okay. I'm an adventurous eater. I'll try anything once. But the fermented horse milk was probably the least favorite dish/drink I have ever had in all of my travels." — **Kit Reyes**

#13: Pufferfish (Fugu) in Japan

#14: Minke whale in Iceland

"I definitely tried the Icelandic dishes including the shark, the minke whale, the monkfish, the langoustine, the puffin bird, and the Skyr yogurt tasting dairy, and the ice cream made of rye bread!"— **Dondon**

#15: Barbequed insects in Thailand and Vietnam

#16: Kangaroo pie in Australia

#17: Zebra steak in Mombasa, Kenya

#18: Buffalo and ostrich meat in Kenya

#19: Cooked penises of a variety of animals in Beijing, China

#20: Soup Number Five in Philippines

Favorite Foodie Locations

#1: Japan

"I love Japanese food and obviously I gravitate towards Japan where you can't get wrong at any ramen joints. Japanese street-food is underrated and the best place to sample them all is at Dotonburi in Osaka." — **Jon**

#2: Thailand

#3: Spain (especially in Barcelona and San Sebastian)

"I will always remember my first time dining in a three Michelin star restaurant at Akelarre in San Sebastian. The food was out of this world!" — **Dondon**

#4: Italy (especially in Sicily)

"It's a tie between Barcelona and Sicily. We've been to Barcelona countless times and every time, we find a new food spot. Spanish cuisine is to die for and Barcelona offers you a great variety! Sicily is also close to our hearts. From local faves like Arancini to Italian staples like Pasta Aglio e Olio, Sicily is a foodies dream!" — **Rambi**

#5: Taiwan

Highly Recommended Travel Adventure Activities

❑ Zodiac cruise in Antarctica

❑ River cruise in Amazon jungle

❑ Norwegian fjords cruise

❑ Hot air balloon ride in Cappadocia & Bagan

❑ Helicopter diving at Great Barrier Reef & heli-ride at Grand Canyon

❑ Hiking Himalayas, Inca Trail, Kokoda Trail, Mt. Yasur volcano

❑ Ice trek at Perito Moreno and Franz Josef glaciers

❑ Camino de Santiago pilgrimage walk

❑ Safari game drive at Serengeti, Masai Mara, Kruger Park

❑ Arctic and Antarctica polar plunge

❑ Bungee jump and skydive in Queenstown

❑ Cage diving with great white sharks in South Africa

- ❏ Swim with jelly fish in Palau
- ❏ Swim with pigs in the Bahamas
- ❏ Manta ray night snorkel in Hawaii

- ❏ Sailing in San Blas islands and Galapagos
- ❏ Gibbon experience zip line in Laos
- ❏ Stand on the Kjeragbolten rock in Norway

- ❏ Submarine ride in Grand Cayman
- ❏ Trans-Siberian train ride
- ❏ Iron ore train ride in Mauritania

- ❏ Camel ride at Sahara desert
- ❏ Quad bike ride in Namibia
- ❏ Gorilla trek in Uganda, Rwanda

- ❏ High speed driving at the autobahn
- ❏ Death Road mountain bike ride
- ❏ Swag camping in Australian outback

- ❏ Plane spotting at St. Maarten
- ❏ Microlight flight over Victoria Falls
- ❏ Tubing in Vang Vieng

"When I used to frequently go solo backpacking, I had a lack of budget and a desire to see and do EVERYTHING, so adventure would always find itself in the equation. Now that I have a family, I take less risks, but ironically experience more. Some of my more recent adventures would include jumping into open freezing waters during a 'Polar Plunge' in Antarctica and swinging (many times, for dear life) across kilometers of zip lines during the Gibbon experience in Laos." — **Brian**

* * *

"I'll drive a zodiac in Antarctica and go around the iceberg, watch penguins swim, and I'll jump into the water which was too cold!" — **Rinell**

* * *

"I've never ridden a mountain bike before. To do this for the first time at no less than the Death Road, arguably the most dangerous road in the world, was a frightening challenge. Yet I felt I'd regret not doing this top thing for backpackers while I'm already in La Paz, Bolivia. So I went for it: getting acquainted with the bike suspension, being extra careful with every curve to avoid the "babies' heads" fist-sized rocks, and feeling the adrenaline rush of a downhill bike ride that lasted for hours. I was relieved to have survived the ordeal knowing that I have missed being a statistic in recent years where some people have unfortunately did not make it out alive." — **Dondon**

PART 4:

TRAVEL ADVICE

CHAPTER 9

TRAVEL BEHAVIOR

*"When was the last time you did something for
the first time?" – John C. Maxwell*

What travel related question would you ask another traveler?

"What's your cure for jetlag?" — **Kit**

* * *

"How safe is your country?" — **Jazz**

* * *

"Hello, I'm **Odette** — love to travel and write about it; what's your favorite country in terms of culture? Can we pose for a selfie? Or can you take my photo?" — **Odette**

* * *

"What are your tips for traveling through Sub-Saharan Africa? Also, how many countries have you been to?" — **Brian**

* * *

"A question I'd like to ask to those who have completed all 193 United Nations member-countries is this: 'What would you do differently if you were to do the quest all over again?'" — **Riza**

* * *

"How did you do it?" — **Dondon**

* * *

"What city outside of your country would you like to visit again?"
— **Raoul**

* * *

"I would normally ask for his/her age and number of countries and
destinations so far traveled." — **Jimmy**

* * *

"What's the best thing you've seen and the greatest experience from
your travels?" — **Jon**

* * *

"I would ask him/her about the personal interest in traveling; for
instance, if it is culturally motivated, or visiting tribes, festivals and
so on." — **Vhang**

* * *

"Favorite country to visit, to live in, to eat in? Why do you travel?"
— **Rambi**

* * *

"How was it in that place?" — **Badong**

* * *

"What is your favorite place and why?" — **Luisa**

How do you keep yourself entertained while traveling when you're not sightseeing?

"Being a flâneur." — **Kit**

* * *

"Sitting, preferably on a bench, watching people go by. Soon someone will approach me and ask 'Are you alone?' And when I say, 'Yes', then he or she would begin to tell stories. I listen." — **Odette**

* * *

"Browsing the net for my next adventure" — **Badong**

* * *

"Researching on what to do next! Kidding aside, it's discovering local cuisine and getting to know the locals." — **Rambi**

* * *

"Eating local food and watching locals go by. Listening to local music." — **Jazz**

* * *

"Reading guidebooks, watching movies, seeking out local food delicacies, planning itineraries, and chatting with other travelers." — **Brian**

* * *

"Reading a book, catching up with local news, reading up on the local points of interest." — **Raoul**

* * *

"I carry one or two books most of the time which I would read while waiting for someone or my transport or on evenings when I feel like it. I download lots of podcasts when I have internet connection which I listen to when I'm on the road or on the plane." — **Jon**

* * *

"When in the mountains, I look for ways to be mesmerized with the environment. When I'm in the city, I look for a place to drink local alcohol and think on how to plan my future." — **Andie**

* * *

"Looking at unique decorations." — **Jimmy**

* * *

"Talking to local people, going to market and looking for interesting souvenirs." — **Luisa**

* * *

"Arranging my photos and watching some TV series online." — **Vhang**

* * *

"It would involve connecting virtually with my friends and family from wherever I am in the world. I also use the opportunity to bounce off ideas and get suggestions from my travel community. There are a number of helpful communities out there, such as Every Passport Stamp, Travelers' Century Club, Circumnavigators Club, NomadMania, Most Traveled People, Couchsurfing, and Philippine Global Explorers. Some of the active travel Facebook groups I've joined include 'Solo Traveler', 'Worldwide Travel', 'Worldwide Travel Bloggers and Travellers', 'Traveler Insider Bucket List', 'Travel Tips', and 'Add to Bucketlist'. Through my research, I came across online travel forums such as Lonely Planet Thorn Tree, Tripadvisor, Fodor's Travel Talk Forums, Wikitravel, Nomadlist, Travel Massive, r/travel, and r/digitalnomads Reddit, to name a few with a good following."
— **Dondon**

* * *

"When not sightseeing, I'm reading about and planning my next trip or keeping in touch with friends and family through social media. I also do work consultancies and run Explore Africa for Impact and the Philippine Global Explorers remotely." — **Riza**

What is your traveling style (DIY, budget, luxurious, or organized)?

"All of the above! When we were younger, my wife and I would stay at hostels and super cheap accommodations. As we've grown older, we've yearned for more comfortable digs. There were places where we just winged it and there were also places where we've done organized tours. It just depends on the destination!" — **Rambi**

* * *

"DIY — when I didn't have money, I stayed budget at youth hostels or YWCA even in places like New Delhi. Gradually, I climbed up the ladder and stayed in three-star hotels. Then, when I earned more money, I stayed at Paradores (it's a luxury hotel castle and monastery with panoramic views of the monumental cities in Spain).

"Oh, there was one time in Mopti, Mali, Africa, when we were at the Sahara desert, it was almost midnight; there were no street lights and our car broke down. We had no choice but to sleep on a very thin mattress on the cold floor. We just had to be flexible and adjust to the situation." — **Odette**

* * *

"When I first started traveling, I stayed in budget hotels or hostels. If I am traveling solo, I always pick a hostel, particularly a female dorm with ensuite bathroom. If I am traveling with others, I stay in their preferred accommodation: tent, hostel, Airbnb, chain hotel, or luxurious boutique hotel. I've been very lucky to have shared a room in some of the best hotels in the world. Not only have I journeyed with friends, but I've also traveled on small group tours. Recently, I finally found my travel tribe, a group of people who would like to visit

every country in the world or who simply like to go to 'weird' destinations. With the help of the Facebook group 'Every Passport Stamp' and certain bloggers, we've found fixers who have created customized tours for us in destinations as far as Somalia and Syria."
— **April**

* * *

"I do not adhere to a specific travel style. It has been a mix of everything from solo to group tours, DIY to organized tours, ultra-budget to luxurious, economy to business class flights, airport camping to luxury five-star hotels. I want to experience everything as a traveler." —**Riza**

* * *

"My travel style has always been DIY. Note that I organize tours for people too and it's typically mid-range; I may not necessarily pass as a budget traveler type nor am I the luxurious kind. Being a woman, it's safety that has always been my first concern." — **Henna**

* * *

"Depends on the location but I am not fussy…" — **Luisa**

* * *

"I have done many trips on different styles of travel and types of tours, but it's mostly DIY. I love the flexibility of following my own schedule. However, special places require joining organized tours, such as the Machu Picchu Inca Trail, African safaris, or Galapagos." — **Raoul**

> *"A good traveler has no fixed plans and is not intent on arriving."* – Lao Tzu

"I would DIY everywhere whenever possible. I would join an organized tour for off-the-beaten places that are usually dangerous or hard to get to. I did a lot of long-term traveling that lasted anywhere between one month to a year-and-a-half; and since I have time on my hands, I mostly travel by land, taking on the cheapest available public transport and accommodation. When I'm feeling burned-out, I would take a day-off and would go for a staycation at some nice hotel. That said, I am quite flexible but budget conscious." — **Jon**

* * *

"During the early years, it was a low-budget tour; but in later years, I was privileged to taste the luxurious, safe and organized trips especially during my later years." — **Jimmy**

* * *

"My traveling style is definitely DIY and minimalist as much as possible, as I'm always concerned on the allocated budget." — **Andie**

* * *

"Mostly organized by my husband who ranks among the world's top travelers which is certainly an advantage." — **Vhang**

* * *

"In my 20s and early 30s, I was a budget backpacker. Over the past ten years, I have become a more organized and smarter traveler, choosing to spend on some comforts (nicer meal experiences, better lodging) that I would not have paid attention to when I was younger." —**Brian**

* * *

"Relaxed and unorganized, very spontaneous and always off the beaten track." — **Kit**

*   *   *

"Organized tours!" — **Jazz**

*   *   *

"DIY, budget tours!" — **Badong**

Describe some community service activities you were involved in while traveling.

"I wanted to start a new decade in the best way possible, and signed up for a group trip to Bali, Indonesia with Mudita Adventures (formerly, Give Back Give Away). We spent three days volunteering for Bali Children's Project to rebuild, refurbish, and provide books and supplies for an elementary school library in rural Bali. I bawled when the other volunteers and I had finished. My parents were immigrants in the United States, and I knew of the hardship and sacrifice they endured so their children could have different and better opportunities. I acknowledged it as an adult, but seeing the excitement and gratitude on the Indonesian kids' faces as they read their new books overwhelmed me so much because I finally realized how my parents felt each time their sacrifices were rewarded. Of course, the kids and I looked very similar so it was easy to relate how my life could have been different had I been born in the Philippines. I picked Bali as my very first volunteer opportunity, and I wouldn't hesitate to recommend that project or any other ones where Mudita Adventures are involved." — **April**

* * *

"I remember washing dishes in a 24/7 kitchen that fed 100,000 people — free meals, every day, 365 days a year, at the Langar community kitchen in Gurdwara, Amritsar, India. When I was in Bjumbura, Burundi, I visited Les Foyers de Charite which was an orphanage for children aged three-to-twelve. We donated sacks of flour, rice, beans, eggs, several bars of soap, boxes of tooth paste, and gallons of cooking oil. We had a French interpreter and we went back four times, helping children sing songs with laughter." — **Odette**

* * *

"At the beginning of my travels, I used to volunteer in eco-projects and teaching kids English." — **Kach**

* * *

"In Rwanda, I was teaching local children English and also helping to organize their classroom." — **Vhang**

* * *

"I found myself traveling alone in Phnom Penh, Cambodia, close to Christmas Day. On my first night in the city, I went to a local restaurant where, incidentally, an Australian couple was having an engaging conversation about a Christmas project for an orphanage. I overheard them say that they were still looking for volunteers, so I approached them and said I was more than happy to help out. I had nothing else to do anyway and was keen to be part of some sort of yuletide celebration. So on Christmas Day, we went to an orphanage and had to explain the concept of "Christmas" to the kids, which included describing who Santa Claus is. The look in their eyes as they got their gifts from Santa was priceless. After all, it was their first Christmas experience, and I was privileged to witness that."
— **Dondon**

*   *   *

"Visiting children in school and donating books, pencil and candies."
— **Luisa**

*   *   *

"Distributing educational supplies to local school children in Tibet, Nepal, and rural China. And distributing hygiene, medical and educational supplies in Cuba." — **Jazz**

*   *   *

"I did school supplies delivery outreach while hiking amid strong storms, crossing almost waist-deep water rivers in the mountains of General Nakar, Quezon province, Philippines." — **Andie**

*   *   *

"I support grassroots tourism initiatives in local places I visit. I try to use alternative operators who tend to have unique activities like coffee plantation picking-and-process tours in Colombia, which also support the indigenous community." — **Brian**

* * *

"My wife and I partnered with a local charity based in San Andres Bukid, Manila that fed underprivileged kids; we likewise, supported them with their schooling needs. For our daughters' birthdays, we've also asked our friends and family to donate school supplies so that we could share our blessings with these kids. We also adopted a local orphanage for children with special needs in South Korea. We got to play and bond with these special kids every week and we also collected toys and clothes for them for the holidays." — **Rambi**

* * *

"My group did a charity hike on a remote and isolated area in the Sierra Madre Mountains by bringing and installing solar and video players with education videos for the Dumagat local tribes. It was the Dumagat tribe children's first time to see a television set! We also distributed bags and school supplies. There was an intense feeling of happiness; I was so moved that it almost brought me to tears of joy. We walked about four days to reach the place." — **Badong**

* * *

"On my first trip to Cuba, I joined my Canadian travel buddies who ran a canine-and-feline spay/neuter clinic in Havana, Cuba. It was an eye-opener working with the locals in a makeshift clinic that performed routine and surgical procedures for the cats and dogs, strays included. The sense of appreciation and gratitude from the pet owners was a gratifying experience. While many of the locals live on meager resources, I could see their love and concern for their pets. There were stray dogs adopted by state institutions, who work with the local police as patrols — only in Havana!" — **Raoul**

* * *

"Explore Africa For Impact is a social enterprise I founded where all net profits we made in running tours in Africa were funneled back into projects that supported local communities, with priority given to those community service activities that created opportunities for women and girls through education, vocational training, and employment." — **Riza**

What's your favorite souvenir?

"I do not buy souvenirs." — **Riza**

* * *

"Currency and passport stamps — my only collection." — **Kach**

* * *

"Listen to their stories, those are the best souvenirs!" — **Odette**

* * *

341

"Photos I have taken. I rarely buy actual souvenirs." — **Kit**

* * *

"Pictures! I also collect stones or small piece of building material from archeological sites but have stopped when I learned it is illegal." — **Badong**

* * *

"My wife and I used to collect magnets but then realized that the best souvenir for our travels are the pictures we take and the memories we make. Easier on the wallet too!" — **Rambi**

* * *

"I like to collect magnets and T-shirt on every country I go." — **Vhang**

* * *

"I have collected a lot of magnets from all over cause it's easy to carry" — **Luisa**

* * *

"I collect fridge magnets which should indicate the country names with its map shape and flag embedded, souvenir shirts which should have the country names with its flags, and the lowest paper currency. Sometimes I buy local alcohol drinks if it's cheaper in Duty Free" — **Andie**

* * *

"Tea sets!" — **Jazz**

* * *

"I collect a few stuff... I buy Starbucks mugs, without fail at every place I would visit where they have a branch. But they're heavy and bulky so, nowadays, I just buy them when I'm on the last leg of my trip. I collect bills and magnets as well, but there was a phase when I wasn't buying them anymore so my collection isn't complete and extensive. Lately, I've started collecting masks and I have a few really nice ones from West Africa and Indonesia. I only collect those stuff from places that I've traveled to." — **Jon**

* * *

"Hard Rock Café shot glasses!" — **Brian**

* * *

"I collect postcards, Hard Rock Café shot glasses and pins (for friends.) I also have a sizable collection of Hard Rock Café classic t-shirts, as well as Starbucks mugs. Up to some point, I also collected rocks as souvenirs from geologically significant places." — **Raoul**

* * *

"My husband is into fine textiles and artifacts and I also value that. I probably splurge once in a while on jewelry that's a little on the side of refined, archaic ones but not expensive." —**Henna**

* * *

"My favorite souvenir is the Thirteenth (13th) Century Tibetan attire which I purchased in Beijing, China." — **Jimmy**

* * *

"I love to collect miniature replicas of places that I've visited, like a mini Eiffel Tower, Petronas Towers, Space Needle, Golden Gate Bridge, Big Ben, Leaning Tower of Pisa, Statue of Liberty, Blue Mosque, St. Peter's Basilica, Neuschwanstein Castle, and more. However, it was difficult for me to accumulate these souvenirs in my backpack, especially when I was embarking on my twenty-one-month trip around the world, so I needed lightweight souvenirs. So, I started collecting embroidered badges and patches of places I've been which can be sewn into clothes. I ended up with a cool shirt, pair of pants, and a jacket full of embroidered badges, as well as patches of flags and logos of places around the world. Moreover, I started sending postcards to myself, which I sent through my friend's address as all my belongings were in storage since I didn't have a permanent address during my long world trip." — **Dondon**

CHAPTER 10

TRAVEL TIPS & TRICKS

What are some essential traveling skills you used?

"Learning a new language. Spanish comes very handy as it is the main language in South America and you will survive with it even in Portuguese-speaking countries. Basic French, on the other hand, will go a long way in West Africa. During the pandemic, I am learning Bahasa and working on improving my Spanish which would definitely be of use in the long traveling life ahead of me.

Traveling light. When I'm traveling for long periods, I would do a bi-monthly decluttering. If I haven't touched or used a stuff inside my backpack, it means I didn't need it. Traveling light is an essential skill that will save you money, enable you to travel faster, and lessen physical strain.

Diversified money options. Bring cash because your ATM might not work in other countries, credit card for bigger purchases and online bookings, use the ATM machines in more developed countries where the exchange rates are competitive.

Being adept with technologies. There's a lot of value in learning and exploring new apps that continually enhance and optimize traveling. From simple currency converters, offline maps, language translators, flight consolidators, hotel and flight bookings and check-in automations go a long way in making our life as a traveler easier. They're not skills but essential traveling companions. New useful travel-focused applications are coming up all the time but, it is important to be conscientious of their benefit and not sacrifice the social interactions that we would gain if we rely on the one true, proven, and tested formula of getting information: asking the locals." — **Jon**

* * *

"I travel simply if am alone; I am careful to lock the door at night and avoid eating fresh vegetables in some countries." — **Luisa**

* * *

"I am a work-in-progress, always learning on travel skills such as taking pictures of passports, visas, itineraries, and phone numbers." — **Henna**

* * *

"Planning, scheduling flights with the airline apps, including online check-in. In many ways, optimization skills on time allotments, accommodations and costs are a must. On many occasions, I had to decide between cost and value of a particular tour or experience, considering my limited time." — **Raoul**

* * *

"Learn the alphabet, number system, some words to get by. The names of local food and spirits will also help. I memorized the Cyrillic alphabet before I went to Russia so I can read street signs and markings. If the Russian word is just an English word adaptation, then I know what it says. I memorized the way numbers are written and memorized some words to get by like 'hello,' 'where is the bathroom,' 'how much,' etc." — **Jazz**

* * *

"Learn key words of the language of the country you visit. Stay patient. Keep a "glass half-full" perspective: set an expectation that things could go wrong, and if they do, not to freak out, as solutions will emerge." — **Brian**

* * *

"Cross-cultural communication skills are the only one you need, really." — **Kit**

* * *

"Be sure to get a Sim card (minimum amount) for cell phones, in case of emergencies and necessities. Bring small cash enough to exchange at the airport, in case of the need to pay the bus or taxi." — **Odette**

* * *

"Dealing with the local currency, being quick with converting local prices to see if I'm getting a good deal. Being able to understand and speak some Spanish has saved me plenty of times in certain places." — **Rambi**

* * *

> *"Conventional wisdom tells us… we take our baggage with us. I'm not so sure. Travel, at its best, transforms us in ways that aren't always apparent until we're back home. Sometimes we do leave our baggage behind, or, even better, it's misrouted to Cleveland and is never heard from again."* – Eric Weiner

"Virtual tour on area by using Google Tour Creator." — **Badong**

*　*　*

"Monitor the dates of flights with promo fares, haggle for lower charges as reasonably as possible, always be conscious on your surroundings and gut feel; health is very important (ensure that you know yourself; strength, weakness, limits)" — **Andie**

*　*　*

"Traveling can either be a reason to quit your diet, or an exciting opportunity to discover new foods and ways to prepare them. Intermittent fasting makes traveling simpler. It's not a crime to gain weight on vacation - so let yourself live. Like everything in life, balance is a must and your first priority when you're on vacation is to enjoy yourself." — **Auie**

*　*　*

"Being a traveler requires having the skills of a project manager. All considerations around cost, schedule, the scope of destinations, and quality of a trip are factored in. Risks and issues have to be constantly assessed and addressed, such as safety measures and having sufficient resources while traveling, hence there's a lot of assessing of payback and trade-offs, negotiating, and decision making involved. To create a value-for-money travel experience, a traveler must be very resourceful, strategic, tactical, creative, super friendly, persistent, resilient, and entrepreneurial. Those who have become masterful in the art of opportunity scanning and bargain hunting have come to learn some advanced tricks like mock booking, 'fuel dumping', 'skip lagging', and even 'black hat' travel hacking." — **Dondon**

What are the important travel items you always bring with you?

"Two or three passport photos for all-of-a-'sudden immigration rules', e.g. when I drove from Montenegro to enter Albania. Same when entering Swaziland from Maputo, Mozambique. There are times you can give money, but it can lead to serious bribery.

In African countries, always bring the yellow vaccination card.

Two pairs of shoes, in case one gets wet or stolen. New shoes are hard to break-in for comfortable walking." — **Odette**

* * *

"I have a big backpack for my clothes (pack all non-white clothing together to avoid discoloration) and other essentials like tripod and monopod. I try to keep this bag at around 15 to 20 kilograms depending on how much stuff and paper clutter I hoard while on the road. I keep the important things in my small backpack, which includes a waterproof camera and spare SD cards; passport; yellow vaccination card; insurance policy details; credit/debit cards; lightweight laptop; external hard drives to store pictures; chargers and an international adapter; as well as important medicines such as malaria tablets and USANA health supplements; my small teddy bear "Charly" I use to pose for pictures; a watch to keep track of time; and, a mobile phone (which is not really important as long as I have Facebook messenger, Zoom or Whatsapp for communications). I lost two mobile phones (iPhone and Samsung), and I broke five cameras (Canon Powershot S95, S100, and Nikon Coolpix underwater camera) during the course of my first 21-months trip circumnavigating around the world back in 2011." — **Dondon**

* * *

"I learned that I needed to bring some basic medicines, along with my vitamins and health supplements. During one travel instance from Thailand to Vietnam, I got food poisoning along the way. One of the Canadian travelers had antibiotics and offered them to me. I was lucky I got well enough at the end of the tour, to fly back to the United States. Since then, I make it a point to bring basic medicines with me." — **Raoul**

* * *

> *"My dream is to walk around the world. A smallish backpack, all essentials neatly in place. A camera. A notebook. A traveling paint set. A hat. Good shoes. A nice pleated (green?) skirt for the occasional seaside hotel afternoon dance."* — *Maira Kalman*

"I have to make sure I have all the essential medicines." — **Luisa**

* * *

"I always carry two backpacks for months-long trips: One is a 45-to-60-pound backpack for my clothes and pants that should be diverse enough to conform to different climates, toiletries, etc. The other one, a smaller backpack, is for my camera, gadgets and documents. I carry the bigger one on my back and the smaller one in my chest. I only bring one pair of hiking shoes, one which is aesthetically nice-looking, both in the mountains as well as on party places.

Essentials would be: Fuji XT2 mirrorless camera, GoPro, laptop, Bose QC30 headphone, mobile phones, two passports (current and old with valid visas), SIM card holder with phone opener, a good novel, sunglasses, vitamins, and a guidebook." — **Jon**

* * *

"Bose noise-cancelling headphones" — **Jazz**

* * *

"As a Baby Boomer, I still bring along the 'Lonely Planet' guide book, adaptors, cellphones, camera, credit cards and hydro-tooth cleaner." — **Jimmy**

* * *

"Dental floss is an essential travel item for me." — **Riza**

* * *

"My essentials are a book or two to hide money, yes. Most people don't read so they do not bother to touch those. Others are lipstick and toiletries." — **Henna**

* * *

"Quick-dry clothes, lightweight footwear, slippers, travel adapter, an Aeropress Coffee Maker (one of my few indulgences), unlocked cellphone, travel SIM, Swiss army knife, lightweight-yet-sturdy all-weather jacket, airplane travel neck pillow, sunglasses, power bank, rollup water bottle, quick-dry travel towel, two-to-three TSA-approved luggage locks, Alco-gel, ear plugs, portable weighing hook scale, stretchable laundry line, tablet with keyboard, passport, wallet, light backpack, charger and camera." — **Brian**

* * *

"I always travel with a camera. Whether it's a Leica M9 or an iPhone, I always have one with me. I am fascinated by portraiture and documentary photography, and having had a career as a broadcast television and film producer has given me the pretext to travel and document all the places I visit. Being a photographer heightens one's powers of observation when traveling, so photography really makes my journeys feel a lot more vivid and alive. I'm also an extrovert, so this makes getting shots of the people and places I encounter a lot easier." — **Kit**

* * *

"Camera is the most important gadget aside from the passport."
— **Vhang**

*   *   *

"In today's age, my iPhone is a must with all of its travel apps. Comfy shoes are also key, along with track pants or cargo pants that have zippers for pockets. Noice canceling AirPods too. Other than that, I try to travel super light and only bring what is absolutely necessary."
— **Rambi**

*   *   *

"Extra cash, water, extra shirt, power bank, little snack, backpack, phone and a GoPro camera…" — **Badong**

*   *   *

"Hilarity once ensued in Goa, India as I looked for a bathing suit once I realized I forgot to pack one on this trip. I stepped into several shops and couldn't find even one on display in a beach town! It turned out the shopkeepers considered them contraband and would only pull them out behind the counter when asked. From this time onward, I have always packed a bathing suit!

As a pharmacist, I am familiar with the names of generic drugs and can easily buy what I need in a pharmacy in another country. I highly suggest packing electrolyte powder or tablets. If you have a headache, are exhausted, are nursing a hangover, or are experiencing diarrhea, dehydration can be the cause of these ailments. Electrolytes are the fastest commonplace item to combat dehydration.

Unfortunately, I hardly ever find electrolytes in other countries, which is the reason I always pack them.

I am a light packer, and I always carry a reusable bag in case I have to stuff a coat or carry it after I have boarded the plane. I started using this Swiss Gear school bag in late 2018, and I knew if I could travel with winter clothes for three weeks, I would have mastered the art of packing. I officially gave myself that title in October 2019 when I was traveling through the Baltics and 'Stans! Think about this: I managed to pack enough winter clothes to fit in one carry-on bag, and was dressed warmly!

I tend to wear dresses or black jeans for warm climates, and I may switch it up with more leggings and tunic-style shirts when traveling in more conservative countries. In an Islam-dominated country where an *abaya* is highly recommended, I'll pack more leggings. Lastly, when traveling alone in hostels or hotels that seem less secure, I carry a door-stopper to prevent unwanted incidents.

Key notes:

- Roll EVERYTHING or use packing cubes.
- Bring a reusable bag in case you need to stuff your coat or other items on the plane after you have boarded.

- If traveling alone in hotels/hostels that may seem less secure, carry a door-stopper to prevent any unwanted incidents.

The bag should always weigh 7 kg (15 pounds), and I've somehow managed to always bring three pairs of shoes. Ladies, it is totally possible to pack lightly and look fabulous!" — **April**

* April's tips for Women on what to bring in a carry-on bag

- ☐ *1. Quick-dry towel*
- ☐ *2. Reusable bag*
- ☐ *3. Inflatable pillow*
- ☐ *4. Altoids*
- ☐ *5. Adapters*
- ☐ *6. Chargers*
- ☐ *7. Eye mask*
- ☐ *8. Toiletries (TSA approved and easily accessible: shampoo, conditioner, hair cream, body lotion, face lotion, eye cream, toothpaste, mouthwash)*
- ☐ *9. Toiletry bag: bar soap, electric toothbrush, floss, comb, tampons, hair ties, eyebrow tweezer, electrolyte powder, band aids, medicines*
- ☐ *10. Swim: shorts, swimsuit, Cover-Up, goggles, and flip flops*
- ☐ *11. Sleep/work-out: thermal shirt, sports bra, shorts, leggings*
- ☐ *12. 2 to three dresses, 1 skirt, 1 blouse*
- ☐ *13. If destination is Africa: tunic style shirts instead of short dresses*

- [ ] *14. If destination is Islam-dominated: hijab and abaya (three leggings, no dresses)*
- [ ] *15. Gym shoes*
- [ ] *16. 2 pairs of socks*
- [ ] *17. 1 pair of nice flat sandals*
- [ ] *18. 1 door-stopper*
- [ ] *19. 1 bra*
- [ ] *20. 10 pairs of underwear*
- [ ] *21. Folder of papers (specifically for Africa) and stickers*

"In order to achieve my backpacking style, I go with my principle of being practical and go with the minimum carried items, by bringing less materials so that I can fully enjoy the excitement on the road. The following are my own thoughts and based on my experience.

Advantages of Traveling Light:

- Time and budget saver

- Easy and less time to pack, more time on the road

- Easy to buy local souvenir items based on what could only fit in the extra space

- Less strain on back, shoulder or other body parts

- Less worry about leaving items or fitting into storage lockers

- Keeps your mind clear to have enough concentration on the road

- Can move quickly from place to place and it's much comfortable walking on the street / road / trail

- Helps blend in with the locals as traveling light looks normal and ordinary

Disadvantages of Traveling Light:

- Need to invest on very good quality bag that fits your needs which could be costly

- Need to be very cautious on your surroundings as you've got everything to lose in 1 bag

- Could miss useful items due to space constraints (e.g. tripod for night shots, poncho for sudden rains along the trail, jacket for unbearable coldness, party attire for presentable nightlife)."
  — Andie

* Andie's tips for Men on what to bring in a carry-on bag during summer

Note: This is applicable only for the summer season for 1 to 2 days trip. The number of shirts, socks, and underwear is to be increased if the travel days are more than 2 days.

- ☐ *1. Osprey Stratos 34L - breathable mesh back panel, hip belt pockets, internal hydration reservoir, integrated rain cover, weighs 1.3 three kg. I don't bring the personal sling bag anymore.*

- ☐ *2. Trekking socks (wool) — anti-blister, with good cushioning / less sweat, regulates temperature*

- ☐ *3. Dry-fit underwear and Quick Dry towel*

- ☐ *4. Extra Shirt depending on the number of days*

- ☐ *5. Columbia Lightweight trekking pants convertible to shorts*

❑ *6. Medicine Pouch (body pain killer, ointment, paracetamol) — old passports are inside for padding protection of the medicines, I have four passports issued in 2010-2020, three of them already full of stamps.*

❑ *7. Toiletry pouch (toothbrush, toothpaste, bath soap, shampoo, lip balm, perfume, breath freshener) — lip balm / inhaler is inside the hip belt pocket*

❑ *8. Thigh / calve support (for cramps, signs of aging leg muscle pains) — actually these are my savior, my doctors on road, they make me last longer on the trail / street.*

❑ *9. Panasonic Lumix DC-ZS220 zoom-type camera for wildlife photography or breath-taking landscape*

❑ *10. Paper bill pouch for USD cash / paper bill souvenir for give-away — idea came from Hungarian friends during Trans-Siberian train journey, they gave me 500 HUF in exchange of Philippine peso as souvenir. Unfortunately, I didn't have at that time. I included also the other bills in other countries in case someone ask.*

❑ *11. Waterproof Foldable backpack — 15L which can accommodate 1.5 five Liters of water, bubble jacket / camera / selfie stick / gloves.*

❑ *12. Isopropyl Alcohol / Sanitizer / wet wipes — alcohol container should have spray nozzle for insect bites / wound, all of this are inside the front pocket / hip pocket*

❑ *13. Earphone / power bank / charger / cable*

❑ *14. Foldable fan / ball cap / Sandals / ballpen*

❑ *15. Selfie stick, flag buff, sando / shorts for sleeping, reusable plastic for laundry*

* Andie's tips for Men on what to bring in a carry-on bag during winter

Note: This is applicable only for the winter season for 1 to 2 days trip. The number of shirts, socks, and underwear is to be increased if the travel days are more than 2 days.

- ☐ *1. Osprey Atmos 65L (accepted as cabin hand-carry bag) — major advantage of this bag is the removable floating lid that can be replaced with a fixed Flap Jacket. I used this floating lid as my personal bag. I used this bag on my circumnavigation trip from UAE to Uruguay to Chile (with around 6 domestic flights), US and Philippines. This bag was accepted as hand-carry in all of those flights. Same description with the 34Li and weighs 2 kg.*

- ☐ *2. Fleece Gloves / thick gloves*

- ☐ *3. Compression Shirt and pants*

- ☐ *4. Quick Dry towel and dry-fit underwear*

- ☐ *5. Thick Trekking socks (wool) and thick trekking pants*

- ☐ *6. Extra Shirt depending on the number of days*

- ☐ *7. Northface Fleece Jacket / Bubble Jacket— there are lightweight that can withstand -10'C*

- ☐ *8. Bonnet cap with neck cover*

- ☐ *9. Medicine Pouch (Body pain killer, old passports, ointment, paracetamol)*

- ☐ *10. Toiletry pouch (toothbrush, toothpaste, bath soap, shampoo, lip balm, perfume)*

- ☐ *11. Thigh / calve support (for cramps, signs of aging leg muscle pains)*

☐ *12. Panasonic Lumix DC-ZS220 zoom-type camera*

☐ *13. Paper bill pouch for USD cash / paper bill souvenir for give-aways*

☐ *14. Earphone / power bank / charger / cable / ballpen*

☐ *15. Isopropyl Alcohol / Sanitizer/ wet wipes*

☐ *16. Selfie stick*

☐ *17. Flag buff*

☐ *18. Undershirt and shorts for sleeping*

☐ *19. Reusable plastic for laundry*

What are your tips for traveling on a budget?

"Eat in a local place, take a bus, stay in a youth hostel." — **Luisa**

*   *   *

"Accommodation is one of the major expenses in travel that can be circumvented so long as one is prepared to stay in cheap hostels, guesthouses and backpacker lodges. Staying in backpacker lodges also give travelers access to local information." — **Riza**

*   *   *

"Eat a heavy breakfast which is free at your hotel. Skip lunch or bring food from breakfast to see more places. Eat a good dinner with the locals at a nice restaurant sampling local food." — **Jazz**

*   *   *

"Plan early, travel midweek on early morning flights. Do not pass up on hostels — they're more popular with younger people, and they also tend to have scheduled activities. For as long as I have a private accommodation, I am okay with hostels with decent accommodations."
— **Raoul**

*   *   *

"The best advice is to plan well in advance (i.e. flight tickets, train tickets, hostels)… and haggle" — **Andie**

*   *   *

"Look for the best bargains in accommodations, flights, and tours. Consider traveling across contiguous locations that offer the most economical options for transport. When I did my circumnavigation around the world, I traveled across adjacent countries by bus or rail, which sometimes come out cheaper than by air. Also, get involved in a global community network of friends such as Couchsurfing, where people can host in their home city for free. I was fortunate to be part of JCI, a worldwide non-governmental organization, which provided me opportunities to attend international events where I got to meet other members from around the world who I became good friends with some of whom generously offered me to stay at their homes or show me around their city." — **Dondon**

*   *   *

"Travel light, buy flights way ahead of time to ensure good rates, take a deep look at Airbnb experiences (as they often provide off-the-beaten-path excursions at very competitive prices), walk everywhere, research the location you're visiting (e.g. free walking tours to get an early lay of the land are a must). Be flexible with both your time and your itinerary." — **Brian**

* * *

"Buy the tickets well in advance and choose the time when the airlines offer discounted flights, mostly not on weekends. Choose hotels where rooms can be refunded, sometimes we go for last-minute deals." — **Vhang**

* * *

"If you're on a budget, my advice is: don't splurge on hotels. They're usually the most forgettable part of your trip, because most of your time there is spent asleep. That's the thing: it's just a place to sleep. You'll spend most of your time outside of the hotel anyway. Unless you're the kind of traveler who only eats in their hotel when traveling rather than eating out. I would rather splurge on a good meal on my destinations. Or a cultural experience. I've stayed in some of the most expensive, most luxurious, and most exclusive hotels and resorts all over the world for work (not on my dime, of course), and frankly, I don't remember anything from them. As long as the bed is comfortable, and you have privacy (I'm sorry, but I'm way too old to still be staying in hostels), and there are no roaches, and if it's not the most expensive part of your trip, then you're all set." — **Kit**

* * *

"Inquire from the city's Travel Info Center for free guided or walking tours. Eat where locals go (e.g. Basmati rice and slices of tandoori chicken only cost $1 and already includes a drink in India). Ride the public bus — it's more fun, but beware of pick-pockets. Stay at youth hostels. When buying metro subway tickets, inquire about transfers and day-passes instead of buying tickets per ride. Should an injury occur, go to the city's Urgency Care for locals, they help tourists too."
— **Odette**

*   *   *

"Prepare thoroughly and book your accommodation and transport in advance. The earlier you manage to arrange these, the cheaper they would be. I don't normally do this because my schedule can be unstable, especially if I'm employed. But when I'm on longer trips or on one of those months-long journeys, I settle the initial trips as early as I can. For places where the cost of living is quite high and eating out tend to be very expensive (e.g. Norway, Iceland, etc.), cook your own food wherever you're staying or buy from the supermarket. Try CouchSurfing if you're comfortable with that arrangement, otherwise go for hostels. Earn as much points on your credit card or other rewards programs and use them on your travels. Pick evening flights so you'll be on the ground exploring on daytime. Similarly, take those long bus/train rides in the evening so you'll save not only accommodation expenses but you'll have more time in-between places." — **Jon**

*   *   *

"Plan ahead. Research on what you want to do and find out if it's cheaper to by online or onsite. Secure your airfare as soon as you can and you can find crazy steals!" — **Rambi**

Top 10 Travel Hacking Tips from Rambi

1. **Have a goal in mind.** Is it First Class flights to Europe? Or maybe a one-week free over-the-water villa in the Maldives? It helps to have a goal!

2. **Avoid cash like the plague!** Use your credit card for all your expenses. Whenever possible, don't use cash! The key is being disciplined and paying off all balances before they accrue interest.

3. **Not all credit cards are created equally.** Take advantage of 3x airfare, 5x dining, 5x groceries points multipliers from credit card companies.

4. **Pick an airline alliance and hotel program and stick to it.** This will help you consolidate your miles and points so that you can redeem for awards faster.

5. **Manufactured spending.** In order to get more points from credit cards, be creative! Offer to take everyone's cash and pay for dinner using your 3x dining credit card. You can also find a good deal and resell that item for a small profit to earn more miles! You can even pay your taxes with your credit cards. The possibilities are endless!

6. **Buy miles/points when it makes sense.** Aside from collecting miles and points through flying or credit card spending, you can also buy points from airlines and hotels whenever they offer promotions. When redeemed for premium airfare or luxury accommodations, these point promotions can provide great value.

7. **Knowledge is Power.** Get familiar with airline award charts, airline routings and hotel redemption categories. Like most things, knowledge is power in travel hacking so educate yourself in the game.

8. **Use technology to your advantage.** Leverage websites and apps to make travel hacking easier. Expertflyer, AwardWallet, FlightMemory, TripIt are just some tools that you can use.

9. **Be flexible.** Unlike paid fares or hotel stays, you might have to be flexible with your destinations and dates with travel hacking. With proper planning though, you can experience flights, hotels and experiences that otherwise would've been too expensive to buy with cash.

10. **Earn and Burn.** Miles and points continue to devaluate over time as airline and hotels increase award redemptions. Make sure you are always looking for options to spend your hard-earned points. It may be nice looking at your high points balance but it's even nicer enjoying that free first-class seat or luxury hotel stay.

How much is your budget for your travels?

"It's hard to answer because my travel style has changed significantly compared to 2013 to now. But when I quit my job, I only had US$5,000 budget for my intended six-month backpacking trip" — **Kach**

*   *   *

"US$5,000.00 per trip" — **Jazz**

*   *   *

"I have been spending on the average US$1,500 per country. I try to stick to this budget." — **Riza**

*   *   *

"No fixed budget, it really depends as to where I want to go." — **Luisa**

*   *   *

"Depending on the country/island/territory I visit, budget can be adjusted. Sometimes, if the tour guide or driver gives me excellent service, I give an excellent tip, too." — **Odette**

*   *   *

"During the initial years I had a budget of US$3,000 for travels to Western Europe and U.S.A. but in my later years, I had to increase it to US$10,000 to have peace of mind for any vicissitudes and for possible purchase of antique items which could be preserved for posterity." — **Jimmy**

*   *   *

"I spent around US$100,000 when I did my first circumnavigation around the world in 2011 for twenty-one months, where I traveled non-stop to seven continents and seventy-five countries. I don't spend much on food, souvenirs, accommodations, and transportation, but I do splurge on paying for experiences. Some really big-ticket items I've spent on that trip included joining the Antarctica expedition; riding the Trans-Siberian train; attending the London Olympics closing ceremonies; dining in a three-Michelin star restaurant in San Sebastian; joining a Contiki trip in New Zealand where I skydived and went bungee jumping; embarking on a Gibbon zip-line experience in the Laos jungle; making cross-country train travels in Switzerland and Norway; flying to Easter island; sailing in Galapagos and San Blas islands; visiting Oktoberfest; attending a Carnival in Rio de Janeiro; exploring an African safari in Serengeti and Kruger National Park; mingling with a great white shark in a submerged cage; and partying in Ibiza and Koh Phangan, among many others. The experiences were well-worth every cent! No regrets." — **Dondon**

* * *

"When I was younger, my backpacking trips were really rough — CouchSurfing, hostels, public transport, and camping. Budget varies from country-to-country, obviously. Sometimes, I mixed it up: hostels for a week then Airbnb in-between and some luxurious hotels every once-in-a-while, especially when I get burned-out. Generally, in Africa I spend US$50 per day inclusive of food. It's around the same budget in Europe. South America and Asia, however, are way cheaper. I usually indulge myself because I may not go back to that place anymore, so I really seek out authentic good food, lots of coffee, and I drink almost every night." — **Jon**

*   *   *

"I used to be a barebones backpacker (US$50 per day, all in), but now with a little more age and desire for comfort (not luxury), maybe US$1,000 per week, all in, is sufficient." — **Brian**

*   *   *

"I really have no planned or set budget, as it depends on the type of tour and travel. But I am a cost-conscious traveler, I make sure that I pay a fair price for the specific accommodation and class of travel." — **Raoul**

*   *   *

"This depends on the destination. I am willing to pay for quality experiences, but I would never consider myself a luxury traveler. But I am not frugal either." — **Kit**

*   *   *

"This really depends on the destination and on the activities. Cheaper countries, like India or Cambodia, we spend like US$30 to US$50 a day on a low budget." — **Vhang**

*   *   *

"Typically, as low as possible, I'm traveling under my allocated budget. I always account for my expenses on a daily basis." — **Andie**

* * *

"I don't really have a budget and a lot of times we just wing it. It obviously depends on the location but what we try to do is splurge on food and experiences instead of material things." — **Rambi**

* * *

"I worked on a cruise ship; more likely I have either not spent more than US$50 or barely nothing at all in every port we docked. It's a DIY or "go-as-tour" escort for our guests in the ship. I remember the most I that spent was €64 on a day tour going to Paris in 2011 where the ship was just two hours away" — **Badong**

How far in advance do you plan your trip?

"One year ahead for flights; one month ahead for itineraries." — **Brian**

* * *

"Six months to a year in advance." — **Jazz**

* * *

"For longer trips (three months or more), I would start planning my route six months in advance. I work on projects which oftentimes get extended so I normally just book and confirm my flight a month before the trip. I plot my route in a spreadsheet and create a folder in my phone where I would paste or write my notes about the places I'm hitting. Basically, these are information regarding accommodation, activities, places to visit, and local restaurants to try. Once on the road, I'm very flexible and, depending on how I like the place, I would add more days or move fast." — **Jon**

* * *

"Three-to-six months of preparation on average, depending on the length of travel. But sometimes, we do some spontaneous trips as well." — **Vhang**

* * *

"The fun begins two months to research, calling the Consulate for up-to-date visa requirements, inquiries from CouchSurfing members about local customary day-to-day activities — I normally do not stay with them. They share in-depth info, then we meet for coffee or lunch. If an airline offers big discount deals, I go at the spur of the moment. Then, I rely on Google information while waiting at airport terminals." — **Odette**

* * *

"During my days when I was working in a cruise ship; all itinerary of our trips were already provided, so I simply search the net for interesting places nearby when we dock." — **Badong**

* * *

"Being in the military, I have to plan my vacations way in advance. For longer travels, I typically plan between six-to-nine months out. For weekend escapades to other parts of Europe, we've done trips where we only decided to go on the day before we flew out!" — **Rambi**

* * *

"Modesty aside, I really do not have much time in preparing for a trip as I am often guided by the schedule of business conferences. I feel guilty every time I would travel without providing any business justification. One setback of these unplanned trips is the inability to avail of cheaper fares." — **Jimmy**

* * *

"It varies for me; anywhere from last minute spur-of-the-moment to a year in advance." — **Riza**

* * *

"Between two weeks to three months…" — **Luisa**

* * *

"It varies. Since I usually embark on two-week or three-week trips, I can only plan in-between travels. Ideally, I usually plan 90% of the logistics at least three weeks before the trip." — **Raoul**

* * *

"Depends on the destination, trip duration and interest requirements, especially if the country requires a visa before entry. But I usually plan more than a month before the trip. Otherwise, I plan just two weeks before the flight, upon approval of my work vacation leave requests."
— **Andie**

*   *   *

"One week!" — **Kit**

Which are the most valuable resources when planning your journey?

"I try not to plan. I want to be surprised." — **Kit**

*   *   *

"Email locals. For instance, I asked a CouchSurfer in Riyadh: 'Are there Filipinos in Riyadh?' 'Ohh my…Plenty.' When I arrived, I made it a point to meet the Pinoys." — **Odette**

*   *   *

"Internet, travel books by Dorling Kindersley, history books of the place, movies about the country, music of the country…" — **Jazz**

*   *   *

"First, for visas, I Google-searched for 'mfa' together with the country. Second, for flight availability, I used Momondo and Google Flights. Third, for hostels, I used Agoda. Fourth, for the last option, local transportation (I used Uber if available, it's cheaper, too)." — **Andie**

* * *

"Google it!" — **Badong**

* * *

"No doubt, Google offers the most useful resource these days; it has replaced travel guidebooks. The discussions in Every Passport Stamp and other lesser-known travel networks provide useful resources for planning off-the-beaten path travels." — **Riza**

* * *

"Maps, googles and tourist board materials..." — **Luisa**

* * *

"TripAdvisor, Kayak.com, Hotels.hotcom, Google Maps..." — **Raoul**

* * *

"Google maps, Kayak/Expedia/booking.com, Airbnb experiences, xe.com, Tripadvisor..." — **Brian**

* * *

"Lonely Planet or other equally good travel guides, complete travel documentation and sufficient travel funds." — **Jimmy**

* * *

"Still good ol' Lonely Planet, but also the NomadMania website with its wide suggestions of places to visit as part of their Series. I have gotten invaluable insights from being part of groups like Travelers' Century Club, Every Passport Stamp, and Philippine Global Explorers. Nothing beats first-hand experiences from these individuals. For booking air travel, I tend to use Skyscanner and choose 'everywhere' as my destination and see where that leads me." — **Rambi**

* * *

"Most of my travel planning was based on the info I got from the internet, such as things to do in a place, as well as other practical things like visa requirements, maps, weather reports, currency exchanges, details from travel blogs, etc. The best tips of places to visit were things I got as word-of-mouth from fellow travelers I met along the way." — **Dondon**

* * *

"I'm a fan of Lonely Planet books and I keep a collection. They're really handy especially when I'm traveling fast as they offer information that otherwise require a substantial amount of time to research.

I am a recent convert of the NomadMania for the listings of places to see categorized into different interests like museums, lighthouses, UNESCO sites, tentative WHS, railways, castles/palaces, modern architectural buildings, festivals, planetariums, etc. I'm also using the Mark O'Travel app for virtual display of the countries and the U.S. states I've been to. The We Wander PH app to track the provinces in the Philippines is also a good tool. I'm a fan of Instagram, especially because I can save photos of a particular place on different folders I specifically created for every country, which I would later refer back when I'm planning." — **Jon**

TRIVIA: According to the Guinness World Records, the fastest circumnavigation by scheduled flights, visiting six continents, is 56 hours and 56 minutes. This was was achieved by Gunnar Garfors, Ronald Haanstra and Erik de Zwart from 31st January to 2nd February 2018.

Another Guiness World Record is having visited 5 continents in one calendar day. This was achieved by Gunnar Garfors, Adrian Butterworth, Thor Mikalsen, and Sondre Moan Mikalsen.

Favorite Travel Resources

- ❑ rome2rio.com
- ❑ greatescape.co
- ❑ iatatravelcentre.com
- ❑ flightradar24.com
- ❑ dollarflightclub.com
- ❑ scottscheapflights.com
- ❑ secretflying.com
- ❑ expertflyer.com
- ❑ skyscanner.com
- ❑ thepointsguy.com
- ❑ awardwallet.com
- ❑ tripit.com
- ❑ trip.com
- ❑ omio.com
- ❑ seat61.com
- ❑ thetrainline.com
- ❑ momondo.com
- ❑ kayak.com
- ❑ expedia.com

- ❑ couchsurfing.com
- ❑ meetup.com
- ❑ airbnb.com
- ❑ kiwi.com
- ❑ agoda.com
- ❑ booking.com
- ❑ hotels.com
- ❑ hostelbookers.com
- ❑ hostelworld.com
- ❑ viator.com
- ❑ tripadvisor.com
- ❑ getyourguide.com
- ❑ klook.com
- ❑ Travello app
- ❑ loungebuddy.com
- ❑ Priority Pass app
- ❑ Grab app
- ❑ Uber app
- ❑ xe.com

Favorite Travel Resources

- ❑ polarsteps.com
- ❑ esplor.io
- ❑ pebblar.com
- ❑ inspirock.com
- ❑ wanderlog.com
- ❑ Every Passport Stamp FB group
- ❑ nomadmania.com
- ❑ nomadlist.com
- ❑ lonelyplanet.com/thorntree
- ❑ travellerspoint.com
- ❑ google translate
- ❑ google maps
- ❑ maps.me
- ❑ Pocket Earth app
- ❑ App in the Air app
- ❑ Mark O'Travel app
- ❑ We Wander PH app
- ❑ Been app
- ❑ flightmemory.com
- ❑ CityMaps2Go app
- ❑ Citymapper app
- ❑ Moovit app
- ❑ Transit app
- ❑ Waze app
- ❑ Glympse app
- ❑ Life360 Family Locator app
- ❑ Find My Friends app
- ❑ Yandex Maps app
- ❑ accuweather.com
- ❑ Zello app
- ❑ Facebook messenger app
- ❑ Whatsapp app
- ❑ Signal app
- ❑ Viber app
- ❑ WeChat app
- ❑ expressvpn.com
- ❑ nordvpn.com
- ❑ surfshark.com

CHAPTER 11

TRAVEL INSIGHTS

What is the best learning you've gained while traveling?

"Traveling has changed me. It inspired me to be humble. You see so little of yourself when you travel. And I think it's also because I incorporate yoga and meditation into my travels." — **Henna**

* * *

"It makes you humble; plus I learned so much about their history and culture." — **Luisa**

* * *

"The ability to piece together the fragmented histories of the world and to unveil the personalities that had played an important role in today's world and what triggered them to do the same." — **Jimmy**

* * *

"Travel taught me to be more patient and always optimistic." — **Vhang**

* * *

"Patience goes a long way, especially knowing that if anything can go wrong, it will. Also, a big smile goes a long way, when learning the local language." — **Raoul**

* * *

"Getting exposed to different cultures — seeing with your raw eyes the situation on the ground, conversing with the locals and getting to know them on a personal level… their life experiences, desires, and aspirations — erase whatever social biases you may have for people from other races and replaced it with a newfound appreciation and understanding of the plight of others." — **Jon**

*   *   *

"The nuances of language. Like Valhalla is God in Norse countries while Bathala is God in Filipino. Very close resemblance; so interesting. The structure of the Apadana in Persepolis and the Acapana in Tiwanaku Bolivia are similar and the words 'Apadana' and 'Acapana' seem to be related. Indonesian word for 'rice' is *ngasi*. Kapampangan word for rice is '*ngasi.*' Indonesian word for house — '*balay*,' Kapampangan word for house — '*bale.*' Of all the people from Luzon, the Kapampangans have a different character and distinct language when there is no body of water or mountain that separates it from other provinces. My hypothesis is Indonesians must have settled in the area of Pampanga; thus, they have a very distinct dialect from all the people in Luzon and a very distinct character." — **Jazz**

*   *   *

"How not to stand out and instead, blend in — my family used to say that whenever I would travel somewhere, I'd come back looking like a local. Learning to be more street-smart and respectful of the local culture has helped me to avoid potentially dangerous situations throughout my travels." — **Brian**

*   *   *

"That we don't have to be scared of what we don't know or haven't been to, and that's it is absolutely okay to be vulnerable." — **Kit**

* * *

"Learning different cultures." — **Badong**

* * *

"For historical information, meeting and greeting the locals is important. For example, I never knew about the White Cowboys of Tibet. I thought they were mostly monks who wear yellow steeple-crowned hats that resemble the shape of a rooster's comb!

I also came to know, through my travels, that Spain has footprints of dinosaurs in the Gijon region — tracks million years' old." — **Odette**

* * *

"In all my classes, one of the things I always stress to my students is that, we just don't travel, we have to be sustainable travelers." — **Ivan**

* * *

"I'm very, very thankful that I have this opportunity to visit these countries knowing where I came from. Backpacking for a limited time and on a budget make the adventure more about survival; the real character within me pushed my limits beyond my wildest imagination. Be positive at all times, train your mind to see the good in everything. And health is very important!" — **Andie**

* * *

"Having a deeper understanding of the human condition — that we are all truly one and the same at our very core despite all of our differences." — **Dondon**

* * *

"Realizing that the world has inherently good people. Just by keeping an open mind, you will meet beautiful and genuine people in all corners of the world!" — **Rambi**

* * *

"The best learning is this: People are good all over the world. There are only a few bad ones. The chances of you coming across one would be the same, if you had traveled the world, or if you had just stayed home." — **Riza**

What were some of your rookie mistakes while traveling?

"Overthinking and over-planning my travels, such as sightseeing bucket lists, for instance. Or having a schedule. Thankfully, I stopped doing those after my first backpacking trip to Europe in 1997. Since then, I've had a very easy-going mentality when it comes to travel. I think this comes with experience though. But the less expectations we have about traveling, the more rewarding it will probably be. Plan with the barest minimum (or, like me, none at all!), listen, be curious, and you'll find things you never realized you were looking for." — **Kit**

* * *

"The need to have complete data and documentation of places to be visited!" — **Jimmy**

* * *

"Over-packing when we were younger. Not researching enough and missing out on things because of it." — **Rambi**

* * *

"Over-packing, forgetting to bring some useful things and leaving stuff in the hotels." —**Luisa**

* * *

"Rookie mistakes are regular happenings. Starting from packing too many or not packing enough. Forgetting essentials when it was right in your eyes the moment you closed that door." — **Henna**

* * *

"In the beginning, I packed too many clothes and shoes. I also brought extra luggage because I was fond of going to flea markets collecting antiques (e.g. old weighing scales of Romania, old embroidered table cloths of Samarkand, cow bells from Monte Perdido, Spain, rooster clocks from Zurich, farming tools from Lalibela, Ethiopia, scythes, manual meat grinders from

Buenos Aires). It became too much — no more space to display in the house. I stopped. Now I travel very light. My luggage weighs under 15 pounds or 7 kilos." — **Odette**

*   *   *

"I brought too many clothes in the beginning of my travels." — **Vhang**

*   *   *

"Doing an overnight camping, carrying things that I don't really need." — **Badong**

*   *   *

"Bringing lots of stuff that I realized later on that I didn't need and ended up donating to locals. I would say I have improved tremendously on my packing abilities.

Not researching enough, especially on the money restrictions of my country destinations. Normally, I would just withdraw cash from the local ATM's upon arrival. I am using my Singapore-bank-issued ATM and didn't know that Russia and Singapore do not have reciprocal financial arrangements, meaning I cannot use my ATM in Russia. Even I have credit cards available, cash is still king around Russia. I immediately had to transfer money to my friends in Singapore and they, in turn, send those over to me via Western Union." — **Jon**

*   *   *

"Not separating items needed for carry-ons for long layovers (e.g. chargers, adapters, etc., spare clothes and toiletries)." — **Raoul**

*   *   *

"Falling for a scam in Bangkok during my first time to travel abroad." — **Andie**

*   *   *

"Apart from putting my passport in the freezer, I have occasionally procrastinated; either I've done my research way too late or failed to do the research at all. This has led to difficulties getting into countries (i.e. forgetting or not preparing my proof of yellow fever immunization), incurring additional expenses (i.e. getting hit by expensive visa-on-arrivals when reaching a country) and encountering hair-raising situations that could have been avoided (i.e. getting into a not-so-safe country at midnight; not researching the taxi situation; not arranging for a pick up beforehand; or Google-mapping a route through a city that unexpectedly goes through the dodgier parts of town)." — **Brian**

*   *   *

"On my very first trip to Europe in 2006, I landed at Malpensa international airport in Milan. One of my biggest rookie mistakes of all time is assuming that there was only one international airport in a city. When I was about to depart from Italy to go back to the Philippines, I couldn't find my flight on the screen. So, imagine

to my shock and horror, when I was told by Customer Service that I was in the wrong international airport! I was supposed to fly out from Linate International Airport, which was around 50 km away from Malpensa, I had less than two hours before departure, and my Schengen visa was expiring that day! I was seriously thinking of hiring a helicopter to get to the other airport but ended up riding the fastest Mercedes Benz taxi limousine service I've ever had. Thankfully, I made it just in the nick of time, but I will never forget my lesson: never assume and always read the details of my ticket itinerary!"
— **Dondon**

What are the things you don't like about traveling?

"I actually hate flying. I am not scared of it but I think of flying as merely getting me from A to B in the most boring and tiring way. I would prefer to drive and see sceneries, people and wildlife along the way. I can never sleep or sit upright on a plane, and the food is awful!"
— **Riza**

*   *   *

"Long flights!" — **Badong**

*   *   *

"The long flights, especially on the return, and then there's the jet lag. Secondly, the long lines at customs and immigration." — **Raoul**

*   *   *

"I do not enjoy flights anymore. I have probably flown over three million flight miles over the last two decades and I am tired of being herded into cattle class, and then sitting in a metal tube breathing recycled air and eating mediocre food for hours on end. I seriously get bored and restless on flights, maybe because I couldn't sleep. Remember when plane travel was romantic and glorious? Not anymore. Now, flying is no longer a memorable part of travel for me, and every single flight wherein I don't get a free upgrade to business class (which is, sadly, almost every flight) is pure misery." — **Kit**

* * *

"Packing and unpacking!" — **Jazz**

* * *

"It is very inconvenient now to go through security." — **Henna**

* * *

"Lack of sleep, expensive places, scammers, strict airport immigration…" — **Andie**

* * *

"Changing money from one country to another and adjusting to the time difference." — **Luisa**

* * *

"Irregular eating times, I think, is a big sacrifice." — **Vhang**

* * *

"How expensive is this hobby? Jetlag? Traffic? But seriously, I love almost everything about travel. Apart from the destination, I enjoy the journey. Falling in line at the airport, riding the local bus, getting lost — they're all part of the experience." — **Rambi**

* * *

"Travel burnt-out. I normally travel for months on end, especially when I'm on a sabbatical and I would experience getting burned-out after several months. When I was younger, I tend to have more extensive itineraries, so I suffer physically; but through the years I've developed mechanisms to avoid that by putting more variety on my schedule. For example, I would spend some days just doing nothing or staying in my accommodation. I also mixed up my accommodation, moving in-between hostels, Airbnb and nicer hotels.

Stereotyping. There are two examples of this. There are places that by being a Filipino, I am treated in a rather peculiar manner. For example, in both times that I visited Panama, the immigration officer didn't believe that I was merely a tourist and kept insisting that I produce a seafarer's card. The same thing happened to me in Brazil. Second, as dark as I already am, I am still considered a 'white man' in many parts of Africa; thus, making me a target for the typical 'tourist-ripping-off' scheme. There were instances where when I would dine in a simple eatery, I would end up getting billed more than what the locals would normally pay for.

Dual pricing for park and museum entrance. Many countries around the world impose different fees for the local and foreign visitors. This is even more apparent when you will take safaris from countries in Africa. I understand the need to fund these developing countries but visiting these them is not cheap; and, as a tourist, we already spent so much for ancillary tourist services. When it comes out on conversations with open-minded locals on those topics — I would ask them how would they feel if they are charged triple than what I would pay as a local when they take an island-hopping tour in the Philippines; or similarly, if they'll visit Disneyland in other parts of the world. I know that I am privileged just to be able to travel but everything that comes out of my pocket were products of several months of sacrifice just to pursue my love for traveling.

Visa inequality. Holding a Philippine passport and having that dream of visiting every country in the world, I have come across these inconveniences countless times. However, I have come to terms with these limitations because this is something beyond my control."
— **Jon**

* * *

"I am not in favor of travel companions who do not believe in promptness and who do not observe agreements." — **Jimmy**

* * *

"Transitioning from being a solo traveler to a family traveler took some getting used to, as priorities and itineraries have to shift, based on how 'dangerous' a place is or how 'age-appropriate' an activity is. Not to mention that the cost of travel doubles, if not more." — **Brian**

* * *

"When a person beside me, all of a sudden, lights a cigarette; or someone puffs vape smoke in my face. Also, when a flight is delayed, I had to run fast and beg to be placed first in the queue: 'May I go in front… of the line? I am late for my connecting flight.' Some allow me, but others gave me a stoic look." — **Odette**

* * *

"What I don't like about traveling is some parts of what happens before and what happens after. I hate saying goodbyes to people and places, so I get this nostalgic post-travel depression after my trips. I experience something opposite of being homesick – I get 'travel-sick' when I'm back home.

Also, I hope one day that travel planning could be done much more conveniently where I don't have to spend so much time scanning through an overwhelming set of unstructured information. I'm hoping that sometime in the future, I can use some form of artificial intelligence that can recommend destinations and itineraries with best deal offers based on a good understanding of my travel profile that matches my trip history and preferences. How cool would that be." — **Dondon**

* * *

"When you have an irrepressible dream, most of your friends and family cannot fathom the number of steps it will take to achieve that goal. They don't understand the amount of work or sacrifice needed for your dream to be realized. They will never understand the unyielding drive and focus required to achieve your dream, and certainly, most traditional Filipinos will never understand unquenchable wanderlust.

Most of my maternal family immigrated to the United States, and focused on creating a safe haven for their families with the intent of achieving a decent social status in one lifetime. My family always stressed the value of traditional education, the importance of a good job, marriage, and children. By not fulfilling your loved ones' obligations or desires, you can certainly feel isolated and alone, with no one to confide in. It can be extremely lonely.

This loneliness can spread into your travel life, and if you're an extreme traveler like me, who has made a lifestyle of traveling every other week, loneliness is a constant companion. It disappears when traveling with friends, especially when I go with my fellow extreme traveler friends, who share similar goals and levels of wanderlust. Loneliness is a constant silent companion when you know your support system back home isn't supportive, and most certainly, when you're traveling alone and there is a definite language barrier without Google Translate.

> *"I see my path but I don't know where it leads. Not knowing where I'm going is what inspires me to travel it."* – Rosalia de Castro

Although many Filipinos may not fully understand this level of wanderlust, many OFWs would relate to these similar feelings of loneliness, considering they have been away from their families and support systems for many months or years at a time without the family back home seeing the true value of the OFW's sacrifice, hard work, and dedication." — **April**

What inspires you to travel?

"Discovering new places, meeting new people, learning, as well as mixing with locals and different nationalities. I'm inspired because every time I discover new places, I feel an overwhelming joy and I can't explain how it becomes so addictive." — **Badong**

* * *

"The foreignness of the place." — **Kit**

* * *

"Being a nature-lover and adventure-seeker, I've never been so excited during the first time I embarked on that destination. Escaping the stress of personal and professional life, the thought of journeying makes me happiest of all." — **Andie**

* * *

"The prospect of creating new experiences, visiting places where historical events occurred, and where ancient civilizations once thrived, and experiencing adventures to meet like-minded travelers is the best for me." — **Raoul**

* * *

"Fellow travelers. Hearing stories from other travelers about their adventures and how those make them happy and inspire me to seek out those places and experiences.

Images. I am a very visual person and seeing beautiful places triggers that desire to see them with my own eyes.

Books. I owe my interest to traveling from books. Early on, I've always been fascinated by maps, geography, flags and places. The visual images from those beautiful places, portraits of other races and cultures besides what I've been exposed to while growing up inspired me dream of seeing them one day.

Social Media. The advent of social media, specifically Facebook and Instagram, have brought attention to many places that were otherwise unknown. I incorporated the use of social media to narrow down places that are of interest to me.

Travel. In every place that I visit, two or more are added on my list. Traveling is a never-ending cycle of pursuing that which makes you happy." — **Jon**

* * *

"It provides directions on what initiatives I would pursue in this earthly life which is to work hard, keep healthy, save money, join noteworthy associations, being good to your family, and knowing that there would be a reward of travel if I would be successful in pursuing these noteworthy pursuits. It is short of saying that every effort exerted or short-term goal fulfilled was realized with the aforethought that if the same would be done well, the prize of travel would soon be realized." — **Jimmy**

* * *

"Apart from wanting to visit the remaining seventy-six countries on the U.N. list over the next nineteen years (as I don't have the time, flexibility and means to just hunker down and do this over a shorter period), I am very site-oriented. I travel to see new sights, have new experiences, try new cuisines, learn from new cultures and revel in the uniqueness of it all." — **Brian**

* * *

"To be on the road of my journey is to discover the culture and hidden gems of the country." — **Vhang**

* * *

"Traveling is educational to me as I read the history, culture, language, and politics of a place before I travel. The world is a classroom, as there are experiences you feel, observations you see, and people you meet that you cannot learn from books." — **Jazz**

* * *

"Geography class in high school. Reading Lonely Planet books and National Geographic magazines. Stories from locals. Week-long festivals, African cuisine customs and traditions." — **Odette**

*   *   *

"I am inspired by my husband. Odd? Yes. He likes geography, he likes traveling, and he makes it possible for me to do both." — **Henna**

*   *   *

"To see other countries, their way of life, cultures and traditions." — **Luisa**

*   *   *

"The reason I do like to travel is my love for culture. I like to experience many, many different cultures." —**Ivan**

*   *   *

"It is the learning I get from how people live their lives in different parts of the world that inspires me to travel. I like to know what makes them sad and happy." — **Riza**

*   *   *

"On a Saturday morning about eleven years ago, I saw a rainbow right by my parents' house on my drive from work. I thought it was weird because I had never seen one before in Chicago. Really. Later that night, my mother died in her bed, surrounded by several family members. Since then, I've always associated rainbows with my mom. I don't search for them, but I've found them in the dodgiest countries, those kinds of countries that everyone (except for my extreme traveler friends) warn me about. When spotted, I take the rainbow as a sign that my mother is watching over me in these active war zones. Spotting one always gives me comfort in knowing I am protected and I am safe. Earlier in January 2020, my friend and I spotted one during a deep conversation about the loss of our parents. It wasn't a singular one, but a double rainbow, almost like my mom and his dad were each watching over us in Syria. I took comfort in spotting this double rainbow, this excessive sign from my mom that life is truly beautiful and worth living." — **April**

* * *

"Just knowing that there is so much to see in this world. That life is fleeting and tomorrow is not guaranteed." — **Rambi**

* * *

"What inspires me to travel is to fulfill my personal goal of traveling to every country in the world and show every Filipina that achieving an impossible dream is doable. I also want to share and continue my mission of meeting and making one person 1% happier every day on my trips." — **Kach**

* * *

"I love traveling because it puts me in a completely different element. I love new experiences, seeing as much as I can. It is such a great way to break up the monotony. Small stuff like figuring out directions and getting by without the language becomes exciting. And ultimately, nothing quite matches the feeling of being on your own and knowing that you got nothing to rely on. Everything you can accomplish successfully is a product of your resourcefulness and wits (for the most part). It just gives you the feeling that you are more capable of being independent than you ever really believed." — **Auie**

* * *

"To travel is the most precious gift you can give to yourself." — **Rinell**

* * *

"I believe I have the 'wanderlust' gene that gives me this unquenchable longing for traveling the world. I consider myself a connoisseur of travel experiences. As I'm fully aware of my limited existence, my YOLO (you only live once) mantra kicks in powerfully that my FOMO (fear of missing out) levels go through the roof. I love being immersed in adventures where I'm compelled to learn new things, deal with challenges that help me grow as a person, and have an undying fascination exploring the unknown and the unpredictable. Traveling gives me the opportunity to live many lifetimes. I never had it so good!" — **Dondon**

"But man does not create...he discovers." – Antonio Gaudi

The Author

Donalito Bales Jr. is a leader, project manager, entrepreneur, philanthropist and world traveler. He is a co-founder of the Philippine Global Explorers, a not-for-profit global community of filipino world travelers. Since winning the national essay writing competition on migration by the Philippine Department of Foreign Affairs Commission on Filipinos Overseas in 1997, he has come a long way in maturing his writing style. He has written project management articles for Medium and Seven Consulting. Galà is his first published book.

The Editor

Reginald Yu is a certified public accountant by title and a businessman by profession. Although writing and speaking were never his true passions (playing computer games are), his innate "talents" were regularly engaged to compose speeches, edit and/or author manuscripts, render inspirational talks and conduct seminars for various volunteer and charitable organizations, almost always gratuitously, as his personal way of "paying it forward" to his very good friends. This is one of them.

The Artist

Gram Telen is a professional graphic designer and book layout designer with over a decade of experience in this field. He has worked through more than a hundred book cover designs with utmost regard for details and aesthetics. As one of the most accomplished Filipino freelancers specializing in book cover and interior design in Fiverr, he helps turn the vision inside his clients' imagination into a tangible printed book that they will undoubtedly be proud of. That's exactly what he did here.

Made in the USA
Coppell, TX
19 September 2021